HIGH ENERGY PHYSICS
RESEARCH ADVANCES

High Energy Physics
Research Advances

Thomas P. Harrison

AND

Roberto N. Gonzales

Editors

Nova Science Publishers, Inc.

New York

For permission to use material from this book please contact us:
Telephone 631-231-7269; Fax 631-231-8175
Web Site: http://www.novapublishers.com

NOTICE TO THE READER

The Publisher has taken reasonable care in the preparation of this book, but makes no expressed or implied warranty of any kind and assumes no responsibility for any errors or omissions. No liability is assumed for incidental or consequential damages in connection with or arising out of information contained in this book. The Publisher shall not be liable for any special, consequential, or exemplary damages resulting, in whole or in part, from the readers' use of, or reliance upon, this material. Any parts of this book based on government reports are so indicated and copyright is claimed for those parts to the extent applicable to compilations of such works.

Independent verification should be sought for any data, advice or recommendations contained in this book. In addition, no responsibility is assumed by the publisher for any injury and/or damage to persons or property arising from any methods, products, instructions, ideas or otherwise contained in this publication.

This publication is designed to provide accurate and authoritative information with regard to the subject matter covered herein. It is sold with the clear understanding that the Publisher is not engaged in rendering legal or any other professional services. If legal or any other expert assistance is required, the services of a competent person should be sought. FROM A DECLARATION OF PARTICIPANTS JOINTLY ADOPTED BY A COMMITTEE OF THE AMERICAN BAR ASSOCIATION AND A COMMITTEE OF PUBLISHERS.

LIBRARY OF CONGRESS CATALOGING-IN-PUBLICATION DATA

Harrison, Thomas P.
 High energy physics research advances / Thomas P. Harrison and Roberto N. Gonzales.
     p. cm.
 ISBN 978-1-60456-304-7 (hardcover)
 1. Particles (Nuclear physics)--Research. I. Gonzales, Roberto N. II. Title.
QC793.4.H37   2008
539.7'6--dc22   2008003394

Published by Nova Science Publishers, Inc. ✣ New York

CONTENTS

PREFACE

This new book is devoted to recent and important research results in high energy physics which includes the following areas of theoretical and experimental physics: Collider Physics, Underground and Large Array Physics, Astroparticles, Gauge Field Theories, General Relativity and Gravitation, Mathematical Methods of Physics, Solvable Models, Strong Interactions, Weak Interactions, Quantum Field Theory, Statistical Field Theories, String Theory, Supersymmetry, Duality, Branes.

Chapter 1 considers a class of well motivated supersymmetric models of F-term hybrid inflation (FHI) which can be linked to the supersymmetric grand unification. The predicted scalar spectral index n_s cannot be smaller than 0.97 and can exceed unity including corrections from minimal Supergravity, if the number of e-foldings corresponding to the pivot scale k_*=0.002/Mpc is around 50. These results are marginally consistent with the fitting of the three-year Wilkinson microwave anisotropy probe data by the standard power-law cosmological model with cold dark matter and a cosmological constant. However, n_s can be reduced by applying two mechanisms: (i) The utilization of a quasi-minimal Kähler potential with a convenient choice of a sign and (ii) the restriction of the number of e-foldings that k_* suffered during FHI. In the case (i), the authors investigate the possible reduction of n_s without generating maxima and minima of the potential on the inflationary path. In the case (ii), the additional e-foldings required for solving the horizon and flatness problems can be generated by a subsequent stage of fast-roll [slow-roll] modular inflation realized by a sting modulus which does [does not] acquire effective mass before the onset of modular inflation.

In Chapter 2 the authors consider the sl(3,C) affine Toda model coupled to matter (Dirac spinor) (ATM) and through a gauge fixing procedure they obtain the classical version of the generalized sl(3,C) sine-Gordon model (cGSG) which completely decouples from the Dirac spinors. The GSG models are multifield extensions of the ordinary sine-Gordon model. In the spinor sector the authors are left with Dirac fields coupled to cGSG fields. Based on the equivalence between the U(1) vector and topological currents, which holds in the theory, it is shown the confinement of the spinors inside the solitons and kinks of the cGSG model providing an extended hadron model for "quark" confinement [JHEP0701(2007)027]. Moreover, the solitons and kinks of the generalized sine-Gordon (GSG) model are shown to describe the normal and exotic baryon spectrum of two-dimensional QCD. The GSG model arises in the low-energy effective action of bosonized QCD2 with unequal quark mass parameters [JHEP0703(2007)055]. The GSG potential for three flavors resembles the potential of the effec tive chiral lagrangian proposed by Witten to describe low-energy behavior of four dimensional QCD. Among the attractive features of the GSG model are the variety of soliton and kink type solutions for QCD2 unequal quark mass parameters. Exotic baryons in QCD2 [Ellis and Frishman, JHEP0508(2005)081] are discussed in the context of the GSG model. Various semi-classical computations are performed improving previous results and clarifying the role of unequal quark masses. The remarkable double sine Gordon model also arises as a reduced GSG model bearing a kink(K) type solution describing a multi-baryon.

The monopole condensate of five colour QCD is considered in Chapter 3. The näive lowest energy state is unobtainable at one-loop for five or more colours due to simple geometry. The consequent adjustment of the vacuum condensate generates a hierarchy of confinement scales in a natural Higgs-free manner. QCD and QED-like forces emerge naturally, acting upon matter fields that may be interpreted as down quarks, up quarks and electrons.

The authors first apply the transformation of mixing azimuthal and internal coordinate to the 11D M-theory with a stack N M2-branes to find the spacetime of a stack of N D2-branes with Melvin one-form in 10D IIA string theory, after the Kaluza-Klein reduction. Next, they apply the Melvin twist to the spacetime and perform the T duality to obtain the background of a stack of N D3-branes. In the near-horizon limit the background becomes the Melvin field deformed $AdS_5 \times S^5$ with NS-NS B field. In the AdS/CFT correspondence it describes a non-commutative gauge theory with nonconstant non-commutativity in the Melvin field deformed spacetime. The authors study the Wilson loop therein by investigating the classical Nambu-Goto action of the corresponding string configuration and find that, contrast to that in the undeformed spacetime, the string could be localized near the boundary. Our result shows that while the geometry could only modify the Coulomb type potential in IR there presents a minimum distance between the quarks. The authors argue that the mechanism behind producing a minimum distance is coming from geometry of the Melvin field deformed background and is not coming from the space non-commutativity.

As explained in Chapter 5, geometric phase that manifests itself in number of optic and nuclear experiments is shown to be a useful tool for realization of quantum computations in so called holonomic quantum computer model (HQCM). This model is considered as an externally driven quantum system with adiabatic evolution law and finite number of the energy levels. The corresponding evolution operators represent quantum gates of HQCM. The explicit expression for the gates is derived both for one-qubit and for multi-qubit quantum gates as Abelian and non-Abelian geometric phases provided the energy levels to be time-independent or in other words for rotational adiabatic evolution of the system. Application of non-adiabatic geometric-like phases in quantum computations is also discussed for a Caldeira-Legett-type model (one-qubit gates) and for the spin 3/2 quadrupole NMR model (two-qubit gates). Generic quantum gates for these two models are derived. The possibility of construction of the universal quantum gates in both cases is shown.

In Chapter 6 origin of ultra high energy cosmic rays is discussed, namely possible sources, processes in which particles are accelerated, and energetic spectra. The chapter includes five sections. In Section 1 different hypotheses of ultra high energy cosmic rays origin are reported and the list of arrays for their detection is given. From our point of view, cosmic ray sources may be active galactic nuclei. Then it is possible to identify cosmic ray sources directly. The identification procedure is described and results of identification are presented in Section 2. It appears that potential sources are active galactic nuclei with red shifts z<0.01 and Blue Lacertae objects. If this is the case the question is how particles are accelerated to energies more than 10^{20} eV in these objects. This problem is discussed in Section 3. The process of charged particles acceleration in Blue Lacertae objects was suggested elsewhere. Here conditions in active galactic nuclei with moderate luminosities are discussed and possible mechanism of particle acceleration up to 10^{21} eV is considered. In extragalactic space particles interact with cosmic microwave background radiation and therefore unevitably loose energy. As a result a black body (GZK) cut off can appear in the cosmic ray spectrum at ultra high energies. Nearby active nuclei are located at distances less than 40 Mpc (if Hubble constant is 75 km/Mpc·s), and cosmic rays reach the Earth without significant energy losses. Blue Lacertae objects have red shifts from z=0.02 up to z>1 and it is unclear if particles accelerated in these objects can reach the Earth at energies $3 \cdot 10^{20}$ eV, that is the maximal energy detected in cosmic rays. Can our model explain the measured spectrum? Cosmic ray spectra at ultra high energies are considered in Section 4. Spectra of cosmic protons at the Earth are calculated and are compared with the measured one. In addition the limit on maximal cosmic ray energy is derived. In Section 5 main results are listed and predictions of different models are compared.

In: High Energy Physics Research Advances
Editors: T.P. Harrison et al, pp. 1-38

ISBN 978-1-60456-304-7
© 2008 Nova Science Publishers, Inc.

Chapter 1

REDUCING THE SPECTRAL INDEX
IN F-TERM HYBRID INFLATION

C. Pallis[*]

School of Physics and Astronomy, The University of Manchester,
Manchester M13 9PL, UNITED KINGDOM

Abstract

We consider a class of well motivated supersymmetric models of F-term hybrid inflation (FHI) which can be linked to the supersymmetric grand unification. The predicted scalar spectral index n_s cannot be smaller than 0.97 and can exceed unity including corrections from minimal Supergravity, if the number of e-foldings corresponding to the pivot scale $k_* = 0.002/\mathrm{Mpc}$ is around 50. These results are marginally consistent with the fitting of the three-year Wilkinson microwave anisotropy probe data by the standard power-law cosmological model with cold dark matter and a cosmological constant. However, n_s can be reduced by applying two mechanisms: (i) The utilization of a quasi-minimal Kähler potential with a convenient choice of a sign and (ii) the restriction of the number of e-foldings that k_* suffered during FHI. In the case (i), we investigate the possible reduction of n_s without generating maxima and minima of the potential on the inflationary path. In the case (ii), the additional e-foldings required for solving the horizon and flatness problems can be generated by a subsequent stage of fast-roll [slow-roll] modular inflation realized by a sting modulus which does [does not] acquire effective mass before the onset of modular inflation.

1. Introduction

A plethora of precise cosmological observations on the *cosmic microwave background radiation* (CMB) and the large-scale structure in the universe has strongly favored the idea of inflation [1] (for reviews see e.g. Refs. [2, 3, 4]). We focus on a set of well-motivated, popular and quite natural models [5] of *supersymmetric* (SUSY) *F-term hybrid inflation* (FHI) [6], realized [7] at (or very close to) the SUSY *grand unified theory* (GUT) scale $M_{\mathrm{GUT}} = 2.86 \times 10^{16}$ GeV. Namely, we consider the standard [7], shifted [8] and smooth

---

[*]E-mail address: kpallis@auth.gr

[9] FHI. In the context of global SUSY (and under the assumption that the problems of the *standard big bag cosmology* (SBB) are resolved exclusively by FHI), these models predict scalar spectral index, n_s, extremely close to unity and without much running, a_s. Moreover, corrections induced by *minimal supergravity* (mSUGRA) drive [10] n_s closer to unity or even upper than it.

These predictions are marginally consistent with the fitting of the three-year *Wilkinson microwave anisotropy probe* (WMAP3) results by the standard power-law cosmological model with cold dark matter and a cosmological constant (ΛCDM). Indeed, one obtains [11] that, at the pivot scale $k_* = 0.002/\mathrm{Mpc}$, n_s is to satisfy the following rather narrow range of values:

$$n_s = 0.958 \pm 0.016 \;\Rightarrow\; 0.926 \lesssim n_s \lesssim 0.99 \tag{1}$$

at 95% confidence level with negligible a_s.

A possible resolution of the tension between FHI and the data is suggested in Ref. [12]. There, it is argued that values of n_s between 0.98 and 1 can be made to be compatible with the data by taking into account a sub-dominant contribution to the curvature perturbation in the universe due to cosmic strings which may be (but are not necessarily [13]) formed during the phase transition at the end of FHI. However, in such a case, the GUT scale is constrained to values well below M_{GUT} [14, 15, 16]. In the following, we reconsider two other resolutions of the problem above without the existence of cosmic strings:

(i) FHI within *quasi-canonical SUGRA* (qSUGRA). In this scenario, we invoke [16, 17] a departure from mSUGRA, utilizing a quasi-canonical (we use the term coined originally in Ref. [18]) Kähler potential with a convenient arrangement of the sign of the next-to-minimal term. This yields a negative mass term for the inflaton in the inflationary potential which can lead to acceptable n_s's. In a sizable portion of the region in Eq. (1) a local minimum and maximum appear in the inflationary trajectory, thereby jeopardizing the attainment of FHI. In that case, we are obliged to assume suitable initial conditions, so that hilltop inflation [19] takes place as the inflaton rolls from the maximum down to smaller values. Therefore, n_s can become consistent with Eq. (1) but only at the cost of a mild tuning [16] of the initial conditions. On the other hand, we can show [20, 21] that acceptable n_s's can be obtained even without this minimum-maximum problem.

(ii) FHI followed by *modular inflation* (MI). It is recently proposed [22] that a two-step inflationary set-up can allow acceptable n_s's in the context of FHI models even with canonical Kähler potential. The idea is to constrain the number of e-foldings that k_* suffers during FHI to relatively small values, which reduces n_s to acceptable values. The additional number of e-foldings required for solving the horizon and flatness problems of SBB can be obtained by a second stage of inflation (named [22] complementary inflation) implemented at a lower scale. We can show that MI [23] (for another possibility see Ref. [24]), realized by a string modulus, can play successfully the role of complementary inflation. A key issue of this set-up is the evolution of the modulus before the onset of MI [25, 26]. We single out two cases according to whether or not the modulus acquires effective mass before the commencement of MI. We show that, in the first case, MI is of the slow-roll type and a very mild tuning of the initial value of the modulus is needed in order to obtain solution compatible with a number of constraints. In the second case, the initial value of the modulus can be predicted due to its evolution before MI, and MI turns out to be of the fast-roll [27] type. However, in our minimal set-up, an upper bound on the total number of e-foldings obtained during FHI emerges, which signalizes a new disturbing tuning. Possible ways out of this situation are also proposed.

In this presentation we reexamine the above ideas for the reduction of n_s within FHI, implementing the following improvements:

- In the case (i) we delineate the parametric space of the FHI models with acceptable n_s's maintaining the monotonicity of the inflationary potential and derive analytical expressions which approach fairly our numerical results.

- In the case (ii) we incorporate the *nucleosynthesis* (NS) constraint and we analyze the situation in which the inflaton of MI acquires mass before the onset of MI, under some simplified assumptions [28].

The text is organized as follows: In Sec. 2., we review the basic FHI models and in the following we present the two methods for the reduction of n_s using qSUGRA (Sec. 3.) or constructing a two-step inflationary scenario (Sec. 4.). Our conclusions are summarized in Sec. 5..

2. The FHI Models

We outline the salient features (the superpotential in Sec 2.1., the SUSY potential in Sec. 2.2. and the inflationary potential in Sec. 2.3.) of the basic types of FHI and we present their predictions in Sec. 2.6., calculating a number of observable quantities introduced in Sec. 2.4., within the standard cosmological set-up described in Sec. 2.5..

2.1. The Relevant Superpotential

The F-term hybrid inflation can be realized [5] adopting one of the superpotentials below:

$$W = \begin{cases} \kappa S \left(\bar{\Phi}\Phi - M^2 \right) & \text{for standard FHI,} \\ \kappa S \left(\bar{\Phi}\Phi - M^2 \right) - S \frac{(\bar{\Phi}\Phi)^2}{M_S^2} & \text{for shifted FHI,} \\ S \left(\frac{(\bar{\Phi}\Phi)^2}{M_S^2} - \mu_S^2 \right) & \text{for smooth FHI,} \end{cases} \quad \text{where} \quad (2)$$

- S is a left handed superfield, singlet under a GUT gauge group G,

- $\bar{\Phi}$, Φ is a pair of left handed superfields belonging to non-trivial conjugate representations of G, and reducing its rank by their *vacuum expectation values* (v.e.vs),

- $M_S \sim 5 \times 10^{17}$ GeV is an effective cutoff scale comparable with the string scale,

- κ and M, μ_S ($\sim M_{GUT}$) are parameters which can be made positive by field redefinitions.

The superpotential in Eq. (2) for standard FHI is the most general renormalizable superpotential consistent with a continuous R-symmetry [7] under which

$$S \rightarrow e^{i\alpha} S, \ \bar{\Phi}\Phi \rightarrow \bar{\Phi}\Phi, \ W \rightarrow e^{i\alpha} W. \tag{3}$$

Including in this superpotential the leading non-renormalizable term, one obtains the superpotential of shifted [8] FHI in Eq. (2). Finally, the superpotential of smooth [9] FHI can be produced if we impose an extra Z_2 symmetry under which $\Phi \rightarrow -\Phi$ and, therefore, only even powers of the combination $\bar{\Phi}\Phi$ can be allowed.

2.2. The SUSY Potential

The SUSY potential, V_{SUSY}, extracted (see e.g. ref. [2]) from W in Eq. (2) includes F and D-term contributions. Namely,

$$V_{\mathrm{SUSY}} = V_{\mathrm{F}} + V_{\mathrm{D}}, \ \text{where}$$

- The F-term contribution can be written as:

$$V_{\mathrm{F}} = \begin{cases} \kappa^2 M^4 \left((\Phi^2 - 1)^2 + 2S^2\Phi^2 \right) & \text{for standard FHI,} \\ \kappa^2 M^4 \left((\Phi^2 - 1 - \xi\Phi^4)^2 + 2S^2\Phi^2(1 - 2\xi\Phi^2)^2 \right) & \text{for shifted FHI,} \\ \mu_{\mathrm{S}}^4 \left((1 - \Phi^4)^2 + 16S^2\Phi^6 \right) & \text{for smooth FHI,} \end{cases} \tag{4}$$

where the scalar components of the superfields are denoted by the same symbols as the corresponding superfields and

$$\begin{cases} \Phi = |\Phi|/M \ \text{and} \ \mathsf{S} = |S|/M & \text{for standard or shifted FHI ,} \\ \Phi = |\Phi|/2\sqrt{\mu_{\mathrm{S}}M_{\mathrm{S}}} \ \text{and} \ \mathsf{S} = |S|/\sqrt{2\mu_{\mathrm{S}}M_{\mathrm{S}}} & \text{for smooth FHI,} \end{cases}$$

with $\xi = M^2/\kappa M_{\mathrm{S}}$ and $1/7.2 < \xi < 1/4$ [8].

In figs. 1, 2 and 3 we present the three dimensional plot of V_{F} versus Φ and S for standard, shifted and smooth FHI, respectively. The inflationary trajectories are also depicted by bold points, whereas the critical points by red/light points.

- The D-term contribution V_{D} vanishes for $|\bar{\Phi}| = |\Phi|$.

Using the derived V_{SUSY}, we can understand that W in Eq. (2) plays a twofold crucial role:

- It leads to the spontaneous breaking of G. Indeed, the vanishing of V_{F} gives the v.e.vs of the fields in the SUSY vacuum. Namely,

$$\langle S \rangle = 0 \ \text{and} \ |\langle\bar{\Phi}\rangle| = |\langle\Phi\rangle| = v_G = \begin{cases} M & \text{for standard FHI,} \\ \dfrac{M\sqrt{1-\sqrt{1-4\xi}}}{\sqrt{2\xi}} & \text{for shifted FHI,} \\ \sqrt{\mu_{\mathrm{S}}M_{\mathrm{S}}} & \text{for smooth FHI} \end{cases} \tag{5}$$

(in the case where $\bar{\Phi}$, Φ are not *Standard Model* (SM) singlets, $\langle\bar{\Phi}\rangle$, $\langle\Phi\rangle$ stand for the v.e.vs of their SM singlet directions). The non-zero value of the v.e.v v_G signalizes the spontaneous breaking of G.

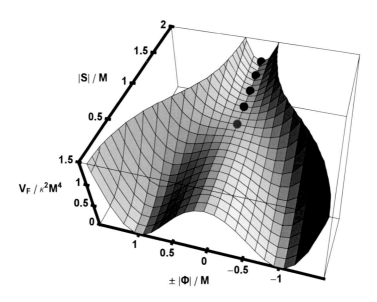

Figure 1. The three dimensional plot of the (dimensionless) F-term potential $V_{\rm F}/\kappa^2 M^4$ for standard FHI versus $S = |S|/M$ and $\pm\,\Phi = \pm|\Phi|/M$. The inflationary trajectory is also depicted by black points whereas the critical point by a red/light point.

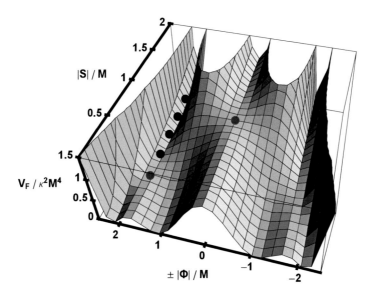

Figure 2. The three dimensional plot of the (dimensionless) F-term potential $V_{\rm F}/\kappa^2 M^4$ for shifted FHI versus $S = |S|/M$ and $\pm\,\Phi = \pm|\Phi|/M$ for $\xi = 1/6$. The (shifted) inflationary trajectory is also depicted by black points whereas the critical points (of the shifted and standard trajectories) are depicted by red/light points.

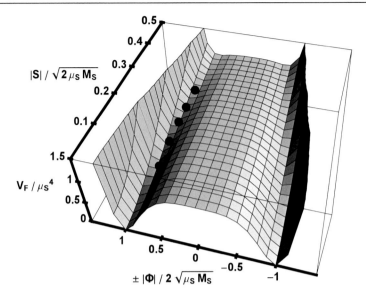

Figure 3. The three dimensional plot of the (dimensionless) F-term potential V_F/μ_S^4 for smooth FHI versus $S = |S|/\sqrt{2\mu_S M_S}$ and $\pm\,\Phi = \pm|\Phi|/2\sqrt{\mu_S M_S}$. The inflationary trajectory is also depicted by black points.

- It gives also rise to FHI. This is due to the fact that, for large enough values of $|S|$, there exist valleys of local minima of the classical potential with constant (or almost constant in the case of smooth FHI) values of V_F. In particular, we can observe that V_F remains practically constant along the following F-flat direction(s):

$$
\begin{array}{ll}
\Phi = 0 & \text{for standard FHI,} \\
\Phi = 0 \ \text{Or} \ \Phi = \sqrt{1/2\xi} & \text{for shifted FHI,} \\
\Phi = 0 \ \text{Or} \ \Phi = 1/2\sqrt{6}S & \text{for smooth FHI.}
\end{array}
$$

From Figs. 1-3 we deduce that the flat direction $\Phi = 0$ corresponds to a minimum of V_F, for $|S| \gg M$, in the cases of standard and shifted FHI and to a maximum of V_F in the case of smooth FHI. Since FHI can be attained along a minimum of V_F we infer that, during standard FHI, the GUT gauge group G is necessarily restored. As a consequence, topological defects such as strings [14, 15, 16], monopoles, or domain walls may be produced [9] via the Kibble mechanism [30] during the spontaneous breaking of G at the end of FHI. This can be avoided in the other two cases, since the form of V_F allows for non-trivial inflationary valleys along which G is spontaneously broken (since the waterfall fields $\bar{\Phi}$ and Φ can acquire non-zero values during FHI). Therefore, no topological defects are produced in these cases. In Table 1 we shortly summarize comparatively the key features of the various versions of FHI.

Table 1. Differences and similarities of the various types of FHI.

	TYPES OF FHI		
	STANDARD	SHIFTED	SMOOTH
The $\Phi = 0$ F-flat direction is:	Minimum for $\|S\| > M$	Minimum for $\|S\| > M$	Maximum
Critical point along the inflationary path?	Yes ($\sigma_c = \sqrt{2}M$)	Yes ($\sigma_c = M_\xi$)	No
Classical flatness of the inflationary path?	Yes	Yes	No
Topological defects?	Yes	No	No

2.3. The Inflationary Potential

The general form of the potential which can drive the various versions of FHI reads

$$V_{\mathrm{HI}} = V_{\mathrm{HI0}} + V_{\mathrm{HIc}} + V_{\mathrm{HIS}} + V_{\mathrm{HIT}}, \quad \text{where:} \tag{6}$$

- V_{HI0} is the dominant (constant) contribution to V_{HI}, which can be written as follows:

$$V_{\mathrm{HI0}} = \begin{cases} \kappa^2 M^4 & \text{for standard FHI,} \\ \kappa^2 M_\xi^4 & \text{for shifted FHI,} \\ \mu_{\mathrm{S}}^4 & \text{for smooth FHI,} \end{cases} \tag{7}$$

with $M_\xi = M\sqrt{1/4\xi - 1}$.

- V_{HIc} is the contribution to V_{HI} which generates a slope along the inflationary valley for driving the inflaton towards the vacua. In the cases of standard [7] and shifted [8] FHI, this slope can be generated by the SUSY breaking on this valley. Indeed, $V_{\mathrm{HI0}} > 0$ breaks SUSY and gives rise to logarithmic radiative corrections to the potential originating from a mass splitting in the $\Phi - \bar{\Phi}$ supermultiplets. On the other hand, in the case of smooth [9] FHI, the inflationary valleys are not classically flat and, thus, there is no need of radiative corrections. Introducing the canonically normalized inflaton field $\sigma = \sqrt{2}|S|$, V_{HIc} can be written as follows:

$$V_{\mathrm{HIc}} = \begin{cases} \frac{\kappa^4 M^4 \mathsf{N}}{32\pi^2}\left(2\ln\frac{\kappa^2 x M^2}{Q^2} + f_c(x)\right) & \text{for standard FHI,} \\ \frac{\kappa^4 M_\xi^4}{16\pi^2}\left(2\ln\frac{\kappa^2 x_\xi M_\xi^2}{Q^2} + f_c(x_\xi)\right) & \text{for shifted FHI,} \\ -2\mu_{\mathrm{s}}^6 M_{\mathrm{S}}^2/27\sigma^4 & \text{for smooth FHI,} \end{cases} \tag{8}$$

with $f_c(x) = (x+1)^2\ln(1+1/x) + (x-1)^2\ln(1-1/x) \Rightarrow f_c(x) \simeq 3$ for $x \gg 1$, $x = \sigma^2/2M^2$ and $x_\xi = \sigma^2/M_\xi^2$. Also N is the dimensionality of the representations to which $\bar{\Phi}$ and Φ belong and Q is a renormalization scale. Although in some parts (see Sec. 4.) of our work rather large κ's are used for standard and shifted FHI, renormalization group effects [31] remain negligible.

In our numerical applications in Secs. 2.6., 3.3., and 4.3. we take N = 2 for standard FHI. This corresponds to the left-right symmetric GUT gauge group $SU(3)_c \times SU(2)_L \times SU(2)_R \times U(1)_{B-L}$ with $\bar{\Phi}$ and Φ belonging to $SU(2)_R$ doublets with $B - L = -1$ and 1 respectively. It is known [13] that no cosmic strings are produced during this realization of standard FHI. As a consequence, we are not obliged to impose extra restrictions on the parameters (as e.g. in Refs. [15, 14]). Let us mention, in passing, that, in the case of shifted [8] FHI, the GUT gauge group is the Pati-Salam group $SU(4)_c \times SU(2)_L \times SU(2)_R$. Needless to say that the case of smooth FHI is independent on the adopted GUT since the inclination of the inflationary path is generated at the classical level and the addition of any radiative correction is expected to be subdominant.

- V_{HIS} is the SUGRA correction to V_{HI}. This emerges if we substitute a specific choice for the Kähler potential K into the SUGRA scalar potential which (without the D-terms) is given by

$$V_{\mathrm{SUGRA}} = e^{K/m_{\mathrm{P}}^2} \left[(F_i)^* K^{i^*j} F_j - 3\frac{|W|^2}{m_{\mathrm{P}}^2} \right], \tag{9}$$

where $F_i = W_i + K_i W/m_{\mathrm{P}}^2$, a subscript i $[i^*]$ denotes derivation *with respect to* (w.r.t) the complex scalar field ϕ^i $[\phi^{i*}]$ and K^{i^*j} is the inverse of the matrix K_{ji^*}. The most elegant, restrictive and highly predictive version of FHI can be obtained, assuming minimal Kähler potential [6, 10], $K_{\mathrm{m}} = |S|^2$. In such a case V_{HIS} becomes

$$V_{\mathrm{HISm}} = V_{\mathrm{HI0}} \frac{\sigma^4}{8m_{\mathrm{P}}^4}, \tag{10}$$

where $m_{\mathrm{P}} \simeq 2.44 \times 10^{18}$ GeV is the reduced Planck scale. We can observe that in this case, no other free parameter is added to the initial set of the free parameters of each model (see Sec. 2.6.).

- V_{HIT} is the most important contribution to V_{HI} from the soft SUSY effects which can be uniformly parameterized as follows [32, 16]:

$$V_{\mathrm{HIT}} = \mathrm{a}_S \sqrt{V_{\mathrm{HI0}}}\,\sigma/\sqrt{2} \tag{11}$$

where a_S is of the order of 1 TeV. V_{HIT} starts [14, 16, 32] playing an important role in the case of standard FHI for $\kappa \lesssim 5 \times 10^{-4}$ and does not have [32], in general, any significant effect in the cases of shifted and smooth FHI.

2.4. Inflationary Observables

Under the assumption that (i) possible deviation from mSUGRA is suppressed (see Sec. 3.2.) and (ii) the cosmological scales leave the horizon during FHI and are not reprocessed during a possible subsequent inflationary stage (see Sec. 4.), we can apply the standard (see e.g. Refs. [2, 3, 4]) calculations for the inflationary observables of FHI. Namely, we can find:

- The number of e-foldings N_{HI*} that the scale k_* suffers during FHI,

$$N_{HI*} = \frac{1}{m_P^2} \int_{\sigma_f}^{\sigma_*} d\sigma \frac{V_{HI}}{V'_{HI}}, \qquad (12)$$

where the prime denotes derivation w.r.t σ, σ_* is the value of σ when the scale k_* crosses outside the horizon of FHI, and σ_f is the value of σ at the end of FHI, which can be found, in the slow roll approximation, from the condition

$$\max\{\epsilon(\sigma_f), |\eta(\sigma_f)|\} = 1, \quad \text{where} \quad \epsilon \simeq \frac{m_P^2}{2}\left(\frac{V'_{HI}}{V_{HI}}\right)^2 \quad \text{and} \quad \eta \simeq m_P^2 \frac{V''_{HI}}{V_{HI}}. \qquad (13)$$

In the cases of standard [7] and shifted [8] FHI and in the parameter space where the terms in Eq. (10) do not play an important role, the end of inflation coincides with the onset of the GUT phase transition, i.e. the slow roll conditions are violated close to the critical point $\sigma_c = \sqrt{2}M$ [$\sigma_c = M_\xi$] for standard [shifted] FHI, where the waterfall regime commences. On the contrary, the end of smooth [9] FHI is not abrupt since the inflationary path is stable w.r.t $\Phi - \bar{\Phi}$ for all σ's and σ_f is found from Eq. (13).

- The power spectrum P_R of the curvature perturbations generated by σ at the pivot scale k_*

$$P_{R*}^{1/2} = \frac{1}{2\sqrt{3}\,\pi m_P^3} \left.\frac{V_{HI}^{3/2}}{|V'_{HI}|}\right|_{\sigma=\sigma_*}. \qquad (14)$$

- The spectral index

$$n_s = 1 + \left.\frac{d\ln P_R}{d\ln k}\right|_{\sigma=\sigma_*} = 1 - m_P^2 \frac{V'_{HI}}{V_{HI}}(\ln P_R)'\bigg|_{\sigma=\sigma_*} = 1 - 6\epsilon_* + 2\eta_*, \qquad (15)$$

and its running

$$\alpha_s = \left.\frac{d^2\ln P_R}{d\ln k^2}\right|_{\sigma=\sigma_*} = \frac{2}{3}\left(4\eta_*^2 - (n_s - 1)^2\right) - 2\xi_*, \qquad (16)$$

where $\xi \simeq m_P^4\, V'_{HI} V'''_{HI}/V_{HI}^2$, the variables with subscript $*$ are evaluated at $\sigma = \sigma_*$ and we have used the identity $d\ln k = H\,dt = -d\sigma/\sqrt{2\epsilon}m_P$.

We can obtain a rather accurate estimation of the expected n_s's if we calculate analytically the integral in Eq. (12) and solve the resulting equation w.r.t σ_*. We pose $\sigma_f = \sigma_c$ for standard and shifted FHI whereas we solve the equation $\eta(\sigma_f) = 1$ for smooth FHI ignoring V_{HIS}. Taking into account that $\epsilon < \eta$ we can extract n_s from Eq. (15). In the case of global SUSY – setting $V_{HIS} = V_{HIT} = 0$ in Eq. (6) – we find

$$n_s = \begin{cases} 1 - 1/N_{HI*} & \text{for standard and shifted FHI,} \\ 1 - 5/3N_{HI*} & \text{for smooth FHI,} \end{cases} \qquad (17)$$

whereas in the context of mSUGRA – setting $V_{\text{HIS}} = V_{\text{HISm}}$ in Eq. (6) – we find

$$n_{\text{s}} = \begin{cases} 1 - 1/N_{\text{HI}*} + 3k^2 N N_{\text{HI}*}/4\pi^2 & \text{for standard FHI,} \\ 1 - 1/N_{\text{HI}*} + 3k^2 N_{\text{HI}*}/2\pi^2 & \text{for shifted FHI,} \\ 1 - 5/3N_{\text{HI}*} + 2\left(6\mu_{\text{S}}^2 M_{\text{S}}^2 N_{\text{HI}*}/m_{\text{P}}^4\right)^{1/3} & \text{for smooth FHI.} \end{cases} \quad (18)$$

Comparing the expressions of Eq. (17) and (18), we can easily infer that mSUGRA elevates significantly n_{s} for relatively large k or M_{S}.

2.5. Observational Constraints

Under the assumption that (i) the contribution in Eq. (14) is solely responsible for the observed curvature perturbation (for an alternative scenario see Ref. [33]) and (ii) there is a conventional cosmological evolution after FHI (see point (ii) below), the parameters of the FHI models can be restricted imposing the following requirements:

(i)   The power spectrum of the curvature perturbations in Eq. (14) is to be confronted with the WMAP3 data [11]:

$$P_{\mathcal{R}*}^{1/2} \simeq 4.86 \times 10^{-5} \text{ at } k_* = 0.002/\text{Mpc}. \quad (19)$$

(ii)   The number of e-foldings N_{tot} required for solving the horizon and flatness problems of SBB is produced exclusively during FHI and is given by

$$N_{\text{tot}} = N_{\text{HI}*} \simeq 22.6 + \frac{2}{3}\ln\frac{V_{\text{HI0}}^{1/4}}{1 \text{ GeV}} + \frac{1}{3}\ln\frac{T_{\text{Hrh}}}{1 \text{ GeV}}, \quad (20)$$

where T_{Hrh} is the reheat temperature after the completion of the FHI.

Indeed, the number of e-foldings N_k between horizon crossing of the observationaly relevant mode k and the end of inflation can be found as follows [2]:

$$\frac{k}{H_0 R_0} = \frac{H_k R_k}{H_0 R_0} = \frac{H_k}{H_0}\frac{R_k}{R_{\text{Hf}}}\frac{R_{\text{Hf}}}{R_{\text{Hrh}}}\frac{R_{\text{Hrh}}}{R_{\text{eq}}}\frac{R_{\text{eq}}}{R_0}$$

$$= \sqrt{\frac{V_{\text{HI0}}}{\rho_{\text{c0}}}}e^{-N_k}\left(\frac{V_{\text{HI0}}}{\rho_{\text{Hrh}}}\right)^{-1/3}\left(\frac{\rho_{\text{Hrh}}}{\rho_{\text{eq}}}\right)^{-1/4}\left(\frac{\rho_{\text{eq}}}{\rho_{\text{m0}}}\right)^{-1/3}$$

$$\Rightarrow N_k \simeq \ln\frac{H_0 R_0}{k} + 24.72 + \frac{2}{3}\ln\frac{V_{\text{HI0}}^{1/4}}{1 \text{ GeV}} + \frac{1}{3}\ln\frac{T_{\text{Hrh}}}{1 \text{ GeV}}. \quad (21)$$

Here, R is the scale factor, $H = \dot{R}/R$ is the Hubble rate, ρ is the energy density and the subscripts 0, k, Hf, Hrh, eq and m denote values at the present (except for the symbols V_{HI0} and $H_{\text{HI0}} = \sqrt{V_{\text{HI0}}}/\sqrt{3}m_{\text{P}}$), at the horizon crossing ($k = R_k H_k$) of the mode k, at the end of FHI, at the end of reheating, at the radiation-matter equidensity point and at the *matter domination* (MD). In our calculation we take into account that $R \propto \rho^{-1/3}$ for *decaying-particle domination* (DPD) or MD and $R \propto \rho^{-1/4}$ for *radiation domination* (RD). We use

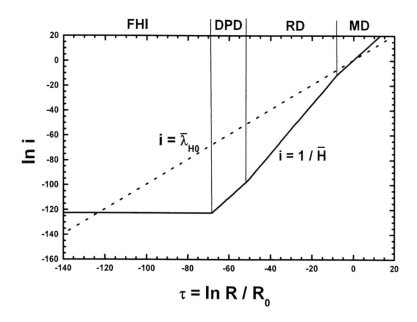

Figure 4. The evolution of the quantities $1/\bar{H} = H/H_0$ with (solid line) and $\bar{\lambda}_{H0} = \lambda/R_0$ (dotted line) as a function of τ for $V_{HI0}^{1/4} = 10^{15}$ GeV, $N_{HI*} \simeq 55$ and $T_{Hrh} = 10^9$ GeV. Shown are also the various eras of the cosmological evolution.

the following numerical values:

$$\rho_{c0} = 8.099 \times 10^{-47} h_0^2 \text{ GeV}^4 \text{ with } h_0 = 0.71,$$

$$\rho_{Hrh} = \frac{\pi^2}{30} g_{\rho*} T_{Hrh}^4 \text{ with } g_{\rho*} = 228.75,$$

$$\rho_{eq} = 2\Omega_{m0}(1 - z_{eq})^3 \rho_{c0} \text{ with } \Omega_{m0} = 0.26 \text{ and } z_{eq} = 3135. \tag{22}$$

Setting $H_0 = 2.37 \times 10^{-4}/\text{Mpc}$ and $k/R_0 = 0.002/\text{Mpc}$ in Eq. (21) we derive Eq. (20).

The cosmological evolution followed in the derivation of Eq. (20) is demonstrated in Fig. 4 where we design the (dimensionless) physical length $\bar{\lambda}_{H0} = \lambda_{H0}/R_0$ (dotted line) corresponding to our present particle horizon and the (dimensionless) particle horizon $\bar{R}_H = 1/\bar{H} = H_0/H$ (solid line) versus the logarithmic time $\tau = \ln R/R_0$. We use $V_{HI0}^{1/4} = 10^{15}$ GeV and $T_{Hrh} = 10^9$ GeV (which result to $N_{HI*} \simeq 55$). We take into account that $\ln \bar{\lambda} \propto \tau$, $\bar{R}_H = H_0/H_{HI0}$ for FHI and $\ln \bar{R}_H \propto 2\tau$ [$\ln \bar{R}_H \propto 1.5\tau$] for RD [MD]. The various eras of the cosmological evolution are also clearly shown.

Fig. 4 visualizes [4] the resolution of the horizon problem of SBB with the use of FHI. Indeed, suppose that $\bar{\lambda}_{H0}$ (which crosses the horizon today, $\bar{\lambda}_{H0}(0) = \bar{R}_H(0)$) indicates the distance between two photons we detect in CMB. In the absence of FHI, the observed homogeneity of CMB remains unexplained since λ_{H0} was outside the horizon, $(\bar{\lambda}_{H0}/\bar{R}_H)(\tau_{LS}) \simeq 33.11$, at the time of *last-scattering* (LS) (with temperature $T_{LS} \simeq 0.26$ eV or logarithmic time $\tau_{LS} \simeq -7$) when the two photons were emitted and so, they could not establish thermodynamic equilibrium. There were 3.6×10^4 disconnected regions within the volume $\bar{\lambda}_{H0}^3(\tau_{LS})$. In other words, photons on the LS surface (with radius

Table 2. Input and output parameters consistent with Eqs. (19) and (20) for shifted ($M_{\rm S} = 5 \times 10^{17}$ GeV) or smooth FHI and $v_G = M_{\rm GUT}$ with and without the mSUGRA contribution.

SHIFTED FHI			SMOOTH FHI		
	WITHOUT	WITH		WITHOUT	WITH
	mSUGRA			mSUGRA	
$\kappa/10^{-3}$	9.2		$M_{\rm S}/5\times10^{17}$ GeV	1.56	0.79
$\sigma_*/10^{16}$ GeV	5.37		$\sigma_*/10^{16}$ GeV	26.8	32.9
$M/10^{16}$ GeV	2.3		$\mu_{\rm S}/10^{16}$ GeV	0.1	0.21
$1/\xi$	4.36		$\sigma_{\rm f}/10^{16}$ GeV	13.4	13.4
$N_{\rm HI*}$	52.2		$N_{\rm HI*}$	52.5	53
$n_{\rm s}$	0.982		$n_{\rm s}$	0.969	1.04
$-\alpha_{\rm s}/10^{-4}$	3.4		$-\alpha_{\rm s}/10^{-4}$	5.8	16.6

$\bar{R}_H(0)$) separated by an angle larger than $\theta = \bar{\lambda}_{\rm LS}(0)/\bar{R}_H(0) \simeq (1/33.11)$ rad $= 1.7^0$ were not in casual contact – here, $\lambda_{\rm LS}$ is the physical length which crossed the horizon at LS. On the contrary, in the presence of FHI, $\lambda_{\rm H0}$ has a chance to be within the horizon again, $\bar{\lambda}_{\rm H0} < \bar{R}_H$, if FHI produces around 56 e-foldings before its termination. If this happens, the homogeneity and the isotropy of CMB can be easily explained as follows: photons that we receive today and were emitted from causally disconnected regions of the LS surface, have the same temperature because they had a chance to communicate to each other before FHI.

2.6. Numerical Results

Our numerical investigation depends on the parameters:

$$\sigma_*, v_G \ \text{and} \ \begin{cases} \kappa & \text{for standard FHI,} \\ \kappa \ \text{with fixed} \ M_{\rm S} = 5 \times 10^{17} \ \text{GeV} & \text{for shifted FHI,} \\ M_{\rm S} & \text{for smooth FHI.} \end{cases}$$

In our computation, we use as input parameters κ or $M_{\rm S}$ and σ_* and we then restrict v_G and σ_* so as Eqs. (19) and (20) are fulfilled. Using Eqs. (15) and (16) we can extract $n_{\rm s}$ and $\alpha_{\rm s}$ respectively which are obviously predictions of each FHI model – without the possibility of fulfilling Eq. (1) by some adjustment.

In the case of standard FHI with N = 2, we present the allowed by Eqs. (19) and (20) values of v_G versus κ (Fig. 5) and $n_{\rm s}$ versus κ (Fig. 6). Dashed [solid] lines indicate results obtained within SUSY [mSUGRA], i.e. by setting $V_{\rm HIS} = V_{\rm HIT} = 0$ [$V_{\rm HIS} = V_{\rm HISm}$ given by Eq. (10) and $V_{\rm HIT}$ given by Eq. (11) with a_S=1 TeV] in Eq. (6). We, thus, can easily identify the regimes where the several contributions to $V_{\rm HI}$ dominate. Namely, for $\kappa \gtrsim 0.01$, $V_{\rm HISm}$ dominates and drives $n_{\rm s}$ to values close to or larger than unity – see Fig. 6. On the other hand, for $5 \times 10^{-4} \lesssim \kappa \lesssim 0.01$, $V_{\rm HIc}$ becomes prominent. Finally, for $\kappa \lesssim 5 \times 10^{-3}$, $V_{\rm HIT}$ starts playing an important role and as v_G increases, $V_{\rm HISm}$ becomes

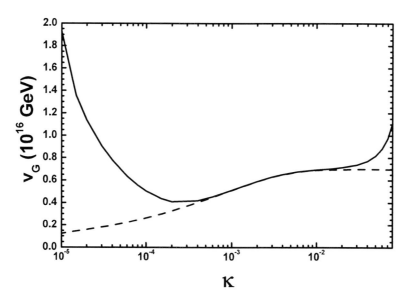

Figure 5. The allowed by Eqs. (19) and (20) values of v_G versus κ for standard FHI with $N = 2$ and $V_{\mathrm{HIS}} = a_S = 0$ (dashed lines) or $V_{\mathrm{HIS}} = V_{\mathrm{HISm}}$ and $a_S = 1$ TeV (solid lines).

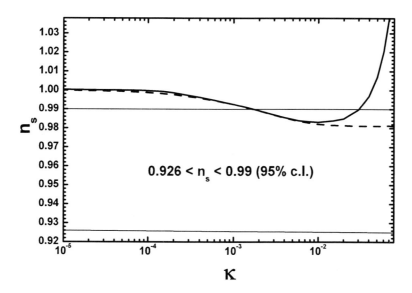

Figure 6. The allowed by Eqs. (19) and (20) values of n_s versus κ for standard FHI with $N = 2$ and $V_{\mathrm{HIS}} = a_S = 0$ (dashed lines) or $V_{\mathrm{HIS}} = V_{\mathrm{HISm}}$ and $a_S = 1$ TeV (solid lines). The region of Eq. (1) is also limited by thin lines.

again important. In Fig. 6 we also design with thin lines the region of Eq. (1). We deduce that there is a marginally allowed area with $0.983 \lesssim n_s \lesssim 0.99$. This occurs for

$$0.0015 \lesssim \kappa \lesssim 0.03 \quad \text{with} \quad 0.56 \lesssim v_G/(10^{16}~\text{GeV}) \lesssim 0.74.$$

in mSUGRA whereas in global SUSY we have $\kappa \gtrsim 0.0015$ and $0.56 \lesssim v_G/(10^{16}~\text{GeV}) \lesssim 0.7$. We realize that $v_G < M_{\text{GUT}}$ – note that $M_{\text{GUT}} = (2 \times 10^{16}/0.7)~\text{GeV}$ where $2 \times 10^{16}~\text{GeV}$ is the mass acquired by the gauge bosons during the SUSY GUT breaking and 0.7 is the unified gauge coupling constant at the scale $2 \times 10^{16}~\text{GeV}$.

In the cases of shifted and smooth FHI we confine ourselves to the values of the parameters which give $v_G = M_{\text{GUT}}$ and display the solutions consistent with Eqs. (19) and (20) in Table 2. We observe that the required κ in the case of shifted FHI is rather low and so, the inclusion of mSUGRA does not raise n_s, which remains within the range of Eq. (1). On the contrary, in the case of smooth FHI, n_s increases sharply within mSUGRA although the result in the absence of mSUGRA is slightly lower than this of shifted FHI. In the former case $|\alpha_s|$ is also considerably enhanced.

3.   Reducing n_s through Quasi-canonical SUGRA

Sizeable variation of n_s in FHI can be achieved by considering a moderate deviation from mSUGRA, named [18] qSUGRA. The form of the relevant Kähler potential for σ is given by

$$K_q = \frac{\sigma^2}{2} \pm c_q \frac{\sigma^4}{4m_P^2} \tag{23}$$

with $c_q > 0$ a free parameter. Note that for $\sigma \ll m_P$ higher order terms in the expansion of Eq. (23) have no effect on the inflationary dynamics. Inserting Eq. (23) into Eq. (9), we obtain the corresponding contribution to V_{HI},

$$V_{\text{HISq}} \simeq V_{\text{HI0}} \left(\mp c_q \frac{\sigma^2}{2m_P^2} + c_{qq} \frac{\sigma^4}{8m_P^4} + \mathcal{O}(\frac{\sigma^6}{m_P}) \right) \quad \text{with} \quad c_{qq} = 1 - \frac{7}{2}c_q + \frac{5}{2}c_q^2. \tag{24}$$

The fitting of WMAP3 data by ΛCDM model obliges [16, 17, 20] us to consider the positive [minus] sign in Eq. (23) [Eq. (24)] (the opposite choice implies [18] a pronounced increase of n_s above unity). As a consequence V_{HI} acquires a rather interesting structure which is studied in Sec. 3.1.. In Sec. 3.2. we specify the observational constraints which we impose to this scenario and in Sec. 3.3. we exhibit our numerical results.

3.1.   The Structure of the Inflationary Potential

In the qSUGRA scenario the potential V_{HI} can be derived from Eq. (6) posing $V_{\text{HIS}} = V_{\text{HISq}}$ given by Eq. (24) with minus in the first term. Depending on the value of c_q, V_{HI} is a monotonic function of σ or develops a local minimum and maximum. The latter case leads to two possible complications: (i) The system gets trapped near the minimum of V_{HI} and, consequently, no FHI takes place and (ii) even if FHI of the so-called hilltop type [19] occurs with σ rolling from the region of the maximum down to smaller values, a mild tuning of the initial conditions is required [16] in order to obtain acceptable n_s's.

It is, therefore, crucial to check if we can accomplish the aim above, avoiding [20, 21] the minimum-maximum structure of $V_{\rm HI}$. In such a case the system can start its slow rolling from any point on the inflationary path without the danger of getting trapped. This can be achieved, if we require that $V_{\rm HI}$ is a monotonically increasing function of σ, i.e. $V'_{\rm HI} > 0$ for any σ or, equivalently,

$$V'_{\rm HI}(\bar{\sigma}_{\rm min}) > 0 \quad \text{with} \quad V''_{\rm HI}(\bar{\sigma}_{\rm min}) = 0 \quad \text{and} \quad V'''_{\rm HI}(\bar{\sigma}_{\rm min}) > 0 \tag{25}$$

where $\bar{\sigma}_{\rm min}$ is the value of σ at which the minimum of $V'_{\rm HI}$ lies. Employing the conditions of Eq. (25) we find approximately:

$$\bar{\sigma}_{\rm min} \simeq \begin{cases} \sqrt{2c_{\rm q}/3c_{\rm qq}}\, m_{\rm P} & \text{for standard and shifted FHI,} \\ \sqrt{2m_{\rm P}/3}\left(\sqrt{5/c_{\rm qq}}\,\mu_{\rm S}M_{\rm S}\right)^{1/4} & \text{for smooth FHI.} \end{cases} \tag{26}$$

Inserting Eq. (26) into Eq. (25), we find that $V_{\rm HI}$ remains monotonic for

$$c_{\rm q} < c_{\rm q}^{\rm max} \quad \text{with} \quad c_{\rm q}^{\rm max} = \begin{cases} 3\kappa\sqrt{c_{\rm qq}{\sf N}}/4\sqrt{2}\pi & \text{for standard FHI,} \\ 3\kappa\sqrt{c_{\rm qq}}/4\pi & \text{for shifted FHI,} \\ (8/3)(c_{\rm qq}/5)^{3/4}\sqrt{\mu_{\rm S}M_{\rm S}}/m_{\rm P} & \text{for smooth FHI.} \end{cases} \tag{27}$$

For $c_{\rm q} > c_{\rm q}^{\rm max}$, $V_{\rm HI}$ reaches at the points $\sigma_{\rm min}$ [$\sigma_{\rm max}$] a local minimum [maximum] which can be estimated as follows:

$$\sigma_{\rm min} \simeq \sqrt{\frac{2c_{\rm q}}{c_{\rm qq}}}\, m_{\rm P} \quad \text{and} \quad \sigma_{\rm max} \simeq \begin{cases} \kappa m_{\rm P}\sqrt{{\sf N}}/2\sqrt{2c_{\rm q}}\pi & \text{for standard FHI,} \\ \kappa m_{\rm P}/2\sqrt{c_{\rm q}}\pi & \text{for shifted FHI,} \\ \sqrt{2/3c_{\rm q}}(\mu_{\rm S}M_{\rm S}m_{\rm P})^{1/3} & \text{for smooth FHI.} \end{cases} \tag{28}$$

Even in this case, the system can always undergo FHI starting at $\sigma < \sigma_{\rm max}$ since $V'_{\rm HI}(\sigma_{\rm max}) = 0$. However, the lower $n_{\rm s}$ we want to obtain, the closer we must set σ_* to $\sigma_{\rm max}$. This signalizes [16] a substantial tuning in the initial conditions of FHI.

Employing the strategy outlined in Sec. (2.4.) we can take a flavor for the expected $n_{\rm s}$'s in the qSUGRA scenario, for any $c_{\rm q}$:

$$n_{\rm s} = \begin{cases} 1 - 2c_{\rm q}\left(1 - 1/c_N\right) - 3c_{\rm qq}\kappa^2{\sf N}c_N/4c_{\rm q}\pi^2 & \text{for standard FHI,} \\ 1 - 2c_{\rm q}\left(1 - 1/c_N\right) - 3c_{\rm qq}\kappa^2 c_N/4c_{\rm q}\pi^2 & \text{for shifted FHI,} \\ 1 - 5/3N_{\rm HI*} + 2\tilde{c}_N - (2\tilde{c}_N N_{\rm HI*} + 7)\,c_{\rm q} & \text{for smooth FHI,} \end{cases} \tag{29}$$

with $c_N = 1 - \sqrt{1 + 4c_{\rm q}N_{\rm HI*}}$ and $\tilde{c}_N = c_{\rm qq}\left(6\mu_{\rm S}^2M_{\rm S}^2 N_{\rm HI*}/m_{\rm P}^4\right)^{1/3}$.

We can clearly appreciate the contribution of a positive $c_{\rm q}$ to the lowering of $n_{\rm s}$.

3.2. Observational Constraints

As in the case of mSUGRA and under the same assumptions, the qSUGRA scenario needs to satisfy Eq. (19) and (20). However, due to the presence of the extra parameter $c_{\rm q}$, a simultaneous fulfillment of Eq. (1) becomes [17, 16, 20] possible. In addition, we take into account, as optional constraint, Eq. (25) so as complications from the appearance of the minimum-maximum structure of $V_{\rm HI}$ are avoided.

Table 3. Input and output parameters consistent with Eqs. (19) and (20) for shifted ($M_S = 5 \times 10^{17}$ GeV) **or smooth FHI,** $v_G = M_{GUT}$ **and selected** n_s**'s within the qSUGRA scenario.**

SHIFTED FHI				SMOOTH FHI			
n_s	0.926	0.958	0.976	n_s	0.926	0.958	0.99
$c_q/10^{-3}$	16.8	7.5	2	$c_q/10^{-3}$	11	8.3	5.45
$c_q^{max}/10^{-3}$	1.7	1.87	2	$c_q^{max}/10^{-3}$	9	9	9
$\sigma_*/10^{16}$ GeV	6.05	5.46	5.36	$\sigma_*/10^{16}$ GeV	23.1	24.5	26.5
$\kappa/10^{-3}$	7.8	8.45	9	$M_S/5 \times 10^{17}$ GeV	2.86	2.02	1.44
$M/10^{16}$ GeV	2.18	2.24	2.28	$\mu_S/10^{16}$ GeV	0.06	0.08	0.1
$1/\xi$	4.1	4.21	4.31	$\sigma_f/10^{16}$ GeV	13.4	13.4	13.4
N_{HI*}	51.7	52	52	N_{HI*}	52.2	52.4	52.6
$-\alpha_s/10^{-4}$	2.8	3.4	3.5	$-\alpha_s/10^{-3}$	0.56	0.8	1

It is worth mentioning that K_q in Eq. (23) generates a non-minimal kinetic term of σ thereby altering, in principle, the inflationary dynamics and the calculation of the inflationary observables. Indeed, the kinetic term of σ is

$$\frac{1}{2}\frac{\partial^2 K_q}{\partial S \partial S_*}\dot{\sigma}^2 \quad \text{with} \quad \frac{\partial^2 K_q}{\partial S \partial S_*} = 1 \pm 2c_q\frac{\sigma^2}{m_P^2} \tag{30}$$

(the dot denotes derivation w.r.t the cosmic time). Assuming that the 'friction' term $3H\dot{\sigma}$ dominates over the other terms in the *equation of motion* (e.o.m) of σ, we can derive the slow roll parameters ϵ and η in Eq. (13) which carry an extra factor $(1 \pm 2c_q\sigma^2/m_P^2)^{-1}$, in the present case. The formulas in Eqs. (12) and (14) get modified also. In particular, a factor $(1 \pm 2c_q\sigma^2/m_P^2)$ must be included in the integrand in the *right-hand side* (r.h.s) of Eq. (12) and a factor $(1 \pm 2c_q\sigma^2/m_P^2)^{1/2}$ in the r.h.s of Eq. (14). However, these modifications are certainly numerically negligible since $\sigma \ll m_P$ and $c_q \ll 1$ (see Sec. 3.3.).

3.3. Numerical Results

Our strategy in the numerical investigation of the qSUGRA scenario is the one described in Sec. 2.6.. In addition to the parameters manipulated there, here we have the parameter c_q which can be adjusted so as to achieve n_s in the range of Eq. (1). We check also the fulfillment of Eq. (25).

In the case of standard FHI with N = 2, we delineate the (lightly gray shaded) region allowed by Eqs. (1), (19) and (20) in the $\kappa - c_q$ (Fig. 7) and $\kappa - v_G$ (Fig. 8) plane. The conventions adopted for the various lines are also shown in the r.h.s of each graphs. In particular, the black solid [dashed] lines correspond to $n_s = 0.99$ [$n_s = 0.926$], whereas the gray solid lines have been obtained by fixing $n_s = 0.958$ – see Eq. (1). The dot-dashed lines correspond to $c_q = c_q^{max}$ in Eq. (27) whereas the dotted line indicates the region in which Eq. (1) is fulfilled in the mSUGRA scenario. In the hatched region, Eq. (25) is also

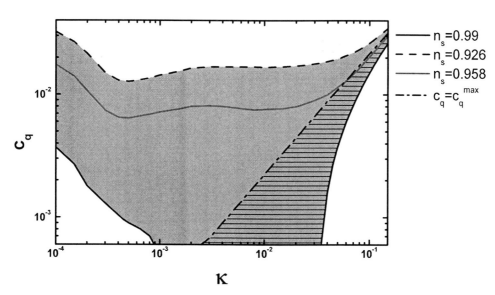

Figure 7. Allowed (lightly gray shaded) region in the $\kappa - c_q$ plane for standard FHI within the qSUGRA scenario. Hatched is the region where the inflationary potential remains monotonic. The conventions adopted for the various lines are also shown.

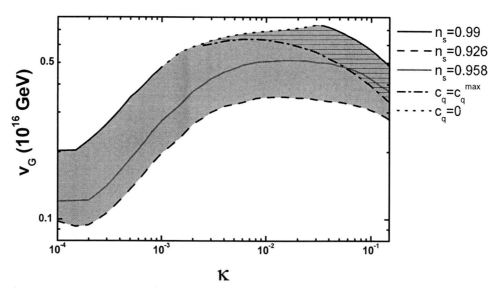

Figure 8. Allowed (lightly gray shaded) region in the $\kappa - v_G$ plane for standard FHI within the qSUGRA scenario. Hatched is the region where the inflationary potential remains monotonic. The conventions adopted for the various lines are also shown.

satisfied. We observe that the optimistic constraint of Eq. (25) can be met in a narrow but not unnaturaly small fraction of the allowed area. Namely, for $n_s = 0.958$, we find

$$0.06 \lesssim \kappa \lesssim 0.15 \text{ with } 0.47 \gtrsim v_G/(10^{16} \text{ GeV}) \gtrsim 0.37 \text{ and } 0.013 \lesssim c_q \lesssim 0.03.$$

The lowest $n_s = 0.946$ can be achieved for $\kappa = 0.15$. Note that the v_G's encountered here are lower that those found in the mSUGRA scenario (see Sec. 2.6.).

In the cases of shifted and smooth FHI we confine ourselves to the values of the parameters which give $v_G = M_{\text{GUT}}$ and display in Table 3 their values which are also consistent with Eqs. (19) and (20) for selected n_s's. In the case of shifted FHI, we observe that (i) it is not possible to obtain $n_s = 0.99$ since the mSUGRA result is lower (see Table 2) (ii) the lowest possible n_s compatible with the conditions of Eq. (25) is 0.976 and so, $n_s = 0.958$ is not consistent with Eq. (25). In the case of smooth FHI, we see that reduction of n_s consistently with Eq. (25) can be achieved for $n_s \gtrsim 0.951$ and so $n_s = 0.958$ can be obtained without complications.

4.   Reducing n_s through a Complementary MI

Another, more drastic and radical, way to circumvent the n_s problem of FHI is the consideration of a double inflationary set-up. This proposition [22] is based on the observation that n_s within FHI models generally decreases [31] with $N_{\text{HI}*}$ – given by Eq. (12). This statement is induced by Eqs. (17) and (18) and can be confirmed by Fig. 9 where we draw n_s in standard FHI with $N = 1$ as a function of $N_{\text{HI}*}$ for several κ's indicated in the graph. On the curves, Eq. (19) is satisfied. Therefore, we could constrain $N_{\text{HI}*}$, fulfilling Eq. (1). Note that a constrained $N_{\text{HI}*}$ was also previously used in Ref. [34] to achieve a sufficient running of n_s.

The residual amount of e-foldings, required for the resolution of the horizon and flatness problems of the standard big-bang cosmology, can be generated during a subsequent stage of MI realized at a lower scale by a string modulus. We show that this scenario can satisfy a number of constraints with more or less natural values of the parameters. Such a construction is also beneficial for MI, since the perturbations of the inflaton field in this model are not sufficiently large to account for the observations, due to the low inflationary energy scale.

Let us also mention that MI naturally assures a low reheat temperature. As a consequence, the gravitino constraint [29] on the reheat temperature of FHI and the potential topological defect problem of standard FHI [30] can be significantly relaxed or completely evaded. On the other hand, for the same reason baryogenesis is made more difficult, since any preexisting baryon asymmetry is diluted by the entropy production during the modulus decay. However, it is not impossible to achieve adequate baryogenesis in the scheme of cold electroweak baryogenesis [35] or in the context of (large) extra dimensions [36].

The main features of MI are sketched in Sec. 4.1.. The parameter space of the present scenario is restricted in Sec. 4.3. taking into account a number of observational requirements which are exhibited in Sec. 4.2.

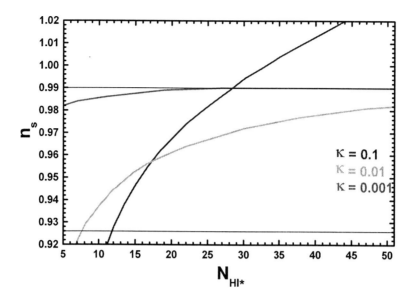

Figure 9. The spectral index n_s in standard FHI as a function of N_{HI*} for several κ's indicated in the graph. On the curves, Eq. (19) is satisfied.

4.1. The Basics of MI

Fields having (mostly Planck scale) suppressed couplings to the SM degrees of freedom and weak scale (non-SUSY) mass are called collectively moduli. After the gravity mediated soft SUSY breaking, their potential can take the form (see the appendix A in Ref. [37]):

$$V_{MI} = (m_{3/2}m_P)^2 \mathcal{V}\left(\frac{s}{m_P}\right) \tag{31}$$

where $\mathcal{V}$ is a function with dimensionless coefficients of order unity and s is the canonically normalized, axionic or radial component of a string modulus. MI is usually supposed [23] to take place near a maximum of V_{MI}, which can be expanded as follows:

$$V_{MI} \simeq V_{MI0} - \frac{1}{2}m_s^2 s^2 + \cdots, \tag{32}$$

where the ellipsis denotes terms which are expected to stabilize V_{MI} at $s \sim m_P$. Comparing Eqs. (31) and (32), we conclude that

$$V_{MI0} = v_s(m_{3/2}m_P)^2 \quad \text{and} \quad m_s \sim m_{3/2}, \tag{33}$$

where $m_{3/2} \sim 1$ TeV is the gravitino mass and the coefficient v_s is of order unity, yielding $V_{MI0}^{1/4} \simeq 3 \times 10^{10}$ GeV. However, if s has just Plank scale suppressed interactions to light degrees of freedom, NS constraint forces us [43] to use (see Sec. 4.2.) much larger values for m_s and $m_{3/2}$. In Fig. 10, we present a typical example of the (dimensionless) potential $V_{MI}/(m_{3/2}m_P)^2$ versus s/m_P, where the constant quantity $c_{MI0} \simeq 0.7$ has been subtracted

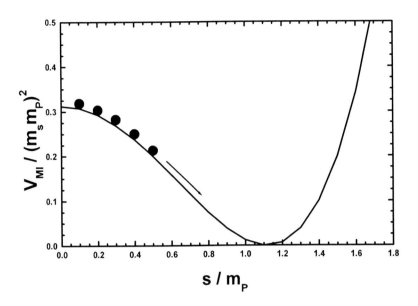

Figure 10.   The (dimensionless) potential $V_{\rm MI}/(m_{3/2}m_{\rm P})^2 = 1 - 0.5(s/m_{\rm P})^2 + 0.2(s/m_{\rm P})^4 - c_{\rm MI0}$ versus $s/m_{\rm P}$. The inflationary trajectory is also depicted by black points.

so that $V_{\rm MI}/(m_{3/2}m_{\rm P})^2$ vanishes at its absolute minimum (the subscript 0 of $V_{\rm MI0}$ and $c_{\rm MI0}$ is not refereed to present-day values).

Solving the e.o.m of the field s (the dot denotes derivation w.r.t the cosmic time),

$$\ddot{s} + 3Hs + d^2V/ds^2 = 0,\qquad(34)$$

for $H = H_s \simeq \sqrt{V_{\rm MI0}}/\sqrt{3}m_{\rm P}$ and $V = V_{\rm MI} \Rightarrow d^2V/ds^2 \simeq -m_s^2$, we can extract [27] its evolution during MI:

$$s = s_{\rm Mi}e^{F_s\Delta N_{\rm MI}} \quad\text{with}\quad F_s \equiv \sqrt{\frac{9}{4} + \left(\frac{m_s}{H_s}\right)^2} - \frac{3}{2}.\qquad(35)$$

Here, $s_{\rm Mi}$ is the value of s at the onset of MI and $\Delta N_{\rm MI}$ is the number of the e-foldings obtained from $s = s_{\rm Mi}$ until a given s. For natural MI we need:

$$0.5 \leq v_s \leq 10 \quad\Rightarrow\quad 2.45 \geq m_s/H_s \geq 0.55 \quad\Rightarrow\quad 1.37 \geq F_s \geq 0.097,\qquad(36)$$

where the lower bound bound on v_s comes from the obvious requirement $V_{\rm MI} > 0$.

In this model, inflation can be not only of the slow-roll but also of the fast-roll [27] type. This is, because there is a range of parameters where, although the ϵ-criterion for MI, $\epsilon_s < 1$, is fulfilled, the η-criterion, $\eta_s < 1$, is violated giving rise to fast-roll inflation. Indeed, using its most general form [4], ϵ_s reads:

$$\epsilon_s = -\frac{\dot{H}_{\rm MI}}{H_{\rm MI}^2} = F_s^2\frac{s^2}{2m_{\rm P}^2},\qquad(37)$$

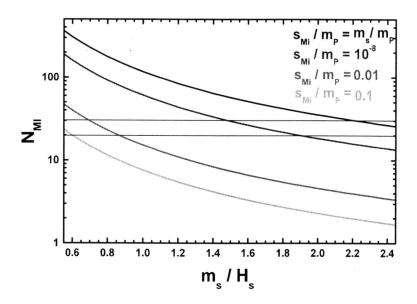

Figure 11. The number of e-foldings N_{MI} obtained during MI as a function of m_s/H_s for $s_{\mathrm{Mf}}/m_{\mathrm{P}} = 1$ and several s_i/m_{P}'s indicated in the graph. The required values of N_{MI} for a successful complementary MI is approximately limited by two thin line.

where the former expression can be derived inserting Eq. (35) into Eq. (34) with $H = H_{\mathrm{MI}} = \sqrt{V_{\mathrm{MI}}}/\sqrt{3}m_{\mathrm{P}}$. Numerically we find:

$$0.005 \leq \epsilon_s \leq 0.94 \quad \text{for} \quad 0.55 \leq m_s/H_s \leq 2.45 \quad \text{and} \quad s/m_{\mathrm{P}} = 1. \tag{38}$$

Therefore, we can obtain accelerated expansion (i.e. inflation) with $H_s \simeq$ cst. Note, though, that near the upper bound on m_s/H_s, ϵ_s gets too close to unity at $s = m_{\mathrm{P}}$ and thus, H_s does not remain constant as s approaches m_{P}. Therefore, our results at large values of m_s/H_s should be considered only as indicative. On the other hand, η_s can be larger or lower than 1, since:

$$|\eta_s| = m_{\mathrm{P}}^2 \frac{|d^2 V_{\mathrm{MI}}/ds^2|}{V_{\mathrm{MI}}} = \frac{m_s^2}{3H_s^2} \simeq \frac{1}{v_s} \tag{39}$$

where the last equality holds for $m_s = m_{3/2}$. Therefore, the condition which discriminates the slow-roll from the fast-roll MI is:

$$\begin{cases} m_s/H_s < \sqrt{3} & \text{or} \quad v_s > 1 \quad \text{for slow-roll MI,} \\ m_s/H_s > \sqrt{3} & \text{or} \quad v_s < 1 \quad \text{for fast-roll MI.} \end{cases} \tag{40}$$

The total number of e-foldings during MI can be found from Eq. (35). Namely,

$$N_{\mathrm{MI}} = \frac{1}{F_s} \ln \frac{s_{\mathrm{Mf}}}{s_{\mathrm{Mi}}} \simeq \frac{1}{F_s} \ln \frac{m_{\mathrm{P}}}{s_{\mathrm{Mi}}}. \tag{41}$$

In our computation we take for the value of s at the end of MI $s_{\mathrm{Mf}} = m_{\mathrm{P}}$, since the condition $\epsilon_s = 1$ gives $s_{\mathrm{Mf}}/m_{\mathrm{P}} = \sqrt{2}/F_s > 1$, for the ranges of Eq. (36). This result is found because the (unspecified) terms in the ellipsis in the r.h.s of Eq. (32) starts playing an important role for $s \sim m_P$ and it is obviously unacceptable.

In Fig. 11, we depict N_{MI} versus m_s/H_s for $s_{Mf} = m_P$ and several s_{Mi}/m_P's indicated in the graph. We observe that N_{MI} is very sensitive to the variations of m_s/H_s. Also, taking into account that $20 \lesssim N_{MI} \lesssim 30$ (limited in Fig. 11 by two thin lines) is needed so that MI plays successfully the role of complementary inflation (see Sec. 4.3.), we can deduce the following:

- As s_{Mi} decreases, the required m_s/H_s for obtaining $N_{MI} \sim 30$ increases. To this end, for $s_{Mi}/m_P \lesssim 10^{-8}$ [$s_{Mi}/m_P \gtrsim 10^{-8}$], we need fast-roll [slow-roll] MI.

- For $s_{Mi}/m_P \gtrsim 0.1$, it is not possible to obtain $N_{MI} \sim 30$ and so, MI can not play successfully the role of complementary inflation.

4.2. Observational Constraints

In addition to Eqs. (1) and (19) – on the assumption that the inflaton perturbation generates exclusively the curvature perturbation – the cosmological scenario under consideration needs to satisfy a number of other constraints too. These can be outlined as follows:

(i) The horizon and flatness problems of SBB can be successfully resolved provided that the scale k_* suffered a certain total number of e-foldings N_{tot}. In the present set-up, N_{tot} consists of two contributions:

$$N_{tot} = N_{HI*} + N_{MI}. \tag{42}$$

Employing the conventions and the strategy we applied in the derivation of Eq. (21), we can find [38] the number of e-foldings N_k between horizon crossing of the observationaly relevant mode k and the end of FHI as follows:

$$
\begin{aligned}
\frac{k}{H_0 R_0} &= \frac{H_k R_k}{H_0 R_0} \\
&= \frac{H_k}{H_0} \frac{R_k}{R_{Hf}} \frac{R_{Hf}}{R_{Mi}} \frac{R_{Mi}}{R_{Mf}} \frac{R_{Mf}}{R_{Mrh}} \frac{R_{Mrh}}{R_{eq}} \frac{R_{eq}}{R_0} \\
&= \sqrt{\frac{V_{HI0}}{\rho_{c0}}} e^{-N_k} \left(\frac{V_{HI0}}{V_{MI0}}\right)^{-1/3} e^{-N_{MI}} \left(\frac{V_{MI0}}{\rho_{Mrh}}\right)^{-1/3} \left(\frac{\rho_{Mrh}}{\rho_{eq}}\right)^{-1/4} \left(\frac{\rho_{eq}}{\rho_{m0}}\right)^{-1/3}
\end{aligned}
$$

$$\Rightarrow \quad N_k + N_{MI} \simeq \ln \frac{H_0 R_0}{k} + 24.72 + \frac{2}{3} \ln \frac{V_{HI0}^{1/4}}{1\ \text{GeV}} + \frac{1}{3} \ln \frac{T_{Mrh}}{1\ \text{GeV}}. \tag{43}$$

Here, we have assumed that the reheat temperature after FHI, T_{Hrh} is lower than $V_{MI0}^{1/4}$ (as in the majority of these models [5]) and, thus, we obtain just MD during the inter-inflationary era. Also, the subscripts Mi, Mf, Mrh denote values at the onset of MI, at the end of MI and at the end of the reheating after the completion of the MI. Inserting into Eq. (43) $H_0 = 2.37 \times 10^{-4}/\text{Mpc}$ and $k/R_0 = 0.002/\text{Mpc}$ and taking into account Eq. (42), we can easily derive the required N_{tot} at k_*:

$$N_{HI*} + N_{MI} \simeq 22.6 + \frac{2}{3} \ln \frac{V_{HI0}^{1/4}}{1\ \text{GeV}} + \frac{1}{3} \ln \frac{T_{Mrh}}{1\ \text{GeV}}. \tag{44}$$

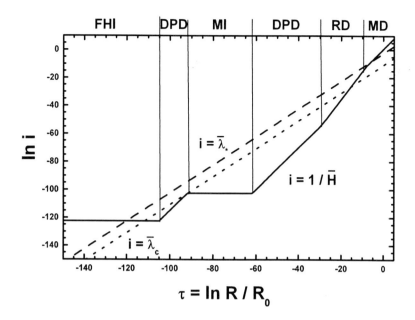

Figure 12. The evolution of the quantities $1/\bar{H} = H_0/H$ (solid line), $\bar{\lambda}_* = \lambda_*/R_0$ (dashed line) and $\bar{\lambda}_c = \lambda_c/R_0$ (dotted line) as a function of τ for $V_{HI0}^{1/4} = 10^{15}$ GeV, $N_{HI*} \simeq 15$, $V_{MI0}^{1/4} = 5 \times 10^{10}$ GeV, $N_{MI} \simeq 30$ and $T_{Mrh} = 1$ GeV. Shown are also the various eras of the cosmological evolution.

The cosmological evolution followed in the derivation of Eq. (43) is illustrated in Fig. 12 where we design the (dimensionless) physical length $\bar{\lambda}_* = \lambda_*/R_0$ (dashed line) corresponding to k_* and the (dimensionless) particle horizon $\bar{R}_H = 1/\bar{H} = H_0/H$ (solid line) as a function of $\tau = \ln R/R_0$. In this plot we take $V_{HI0}^{1/4} = 10^{15}$ GeV, $N_{HI*} \simeq 15$, $V_{MI0}^{1/4} = 5 \times 10^{10}$ GeV, $N_{MI} \simeq 30$, and $T_{Mrh} = 1$ GeV. We take also $\bar{R}_H = H_0/H_s$ for MI. The various eras of the cosmological evolution are shown (compare with Fig. 4).

(ii)   Taking into account that the range of the cosmological scales which can be probed by the CMB anisotropy is [2] $10^{-4}/\text{Mpc} \leq k \leq 0.1/\text{Mpc}$ (length scales of the order of 10 Mpc are starting to feel nonlinear effects and it is, thus, difficult to constrain [39] primordial density fluctuations on smaller scales) we have to assure that all the cosmological scales:

- Leave the horizon during FHI. This entails:

$$N_{HI*} \gtrsim N_k(k = 0.002/\text{Mpc}) - N_k(k = 0.1/\text{Mpc}) = 3.9 \qquad (45)$$

  which is the number of e-foldings elapsed between the horizon crossing of the pivot scale k_* and the scale $0.1/\text{Mpc}$ during FHI.

- Do not re-enter the horizon before the onset of MI (this would be possible since the scale factor increases during the inter-inflationary MD era [38]). This requires

$N_{\mathrm{HI}*} \gtrsim N_{\mathrm{HIc}}$, where N_{HIc} is number of e-foldings elapsed between the horizon crossing of a wavelength k_c (with corresponding length $\bar{\lambda}_c = \lambda_c/R_0$ depicted by a dotted line in Fig. 12) and the end of FHI. More specifically, k_c is to be such that:

$$1 = \frac{k_c}{H_{\mathrm{Mi}} R_{\mathrm{Mi}}} = \frac{H_c R_c}{H_{\mathrm{Mi}} R_{\mathrm{Hf}}} \frac{R_{\mathrm{Hf}}}{R_{\mathrm{Mi}}} = e^{-N_{\mathrm{HIc}}} \left(\frac{V_{\mathrm{HI0}}}{V_{\mathrm{MI0}}}\right)^{1/6} \Rightarrow N_{\mathrm{HIc}} = \frac{1}{6} \ln \frac{V_{\mathrm{HI0}}}{V_{\mathrm{MI0}}}. \quad (46)$$

Both these requirements can be met if we demand [38]

$$N_{\mathrm{HI}*} \gtrsim N_{\mathrm{HI}*}^{\min} \simeq 3.9 + \frac{1}{6} \ln \frac{V_{\mathrm{HI0}}}{V_{\mathrm{MI0}}}. \quad (47)$$

We expect $N_{\mathrm{HI}*}^{\min} \sim 10$ since $(V_{\mathrm{HI0}}/V_{\mathrm{MI0}})^{1/4} \sim 10^{14}/10^{10} \sim 10^4$ and $\ln(10^{16})/6 \sim 6$.

(iii) As it is well known [31, 34], in the FHI models, $|\alpha_s|$ increases as $N_{\mathrm{HI}*}$ decreases. Therefore, limiting ourselves to $|\alpha_s|$'s consistent with the assumptions of the power-law ΛCDM model, we obtain a lower bound on $N_{\mathrm{HI}*}$. Since, within the cosmological models with running spectral index, $|\alpha_s|$'s of order 0.01 are encountered [11], we impose the following upper bound on $|\alpha_s|$:

$$|\alpha_s| \ll 0.01. \quad (48)$$

(iv) Using the bounds of Eq. (36), we can find the corresponding bounds on N_{MI}. Namely,

$$0.73 \ln \frac{m_{\mathrm{P}}}{s_{\mathrm{Mi}}} \leq N_{\mathrm{MI}} \leq 10.2 \ln \frac{m_{\mathrm{P}}}{s_{\mathrm{Mi}}}. \quad (49)$$

The relevant for our analysis (see Sec. 4.3.) is the lower bound on N_{MI} which is $N_{\mathrm{MI}}^{\min} \sim 3$ for $s_{\mathrm{Mi}}/m_{\mathrm{P}} = 0.01$ or $N_{\mathrm{MI}}^{\min} \sim 25$ for $s_{\mathrm{Mi}} \sim H_s$ and $m_s = m_{3/2} = 1$ TeV.

(v) Restrictions on the parameters can be also imposed from the evolution of the field s before MI. Depending whether s acquires or not effective mass [25, 26] during FHI and the inter-inflationary era, we can distinguish the cases:

- If s does not acquire mass (e.g. if s represents the axionic component of a string modulus or if a specific form for the Kähler potential of s has been adopted), we assume that FHI lasts long enough so that the value of s is completely randomized [40] as a consequence of its quantum fluctuations from FHI. We further require that all the values of s belong to the randomization region, which dictates [40] that

$$V_{\mathrm{MI0}} \leq H_{\mathrm{HI0}}^4 \quad \text{where} \quad H_{\mathrm{HI0}}^2 = V_{\mathrm{HI0}}/3m_{\mathrm{P}}^2. \quad (50)$$

Under these circumstances, all the initial values s_{Mi} of s from zero to m_{P} are equally probable – e.g. the probability to obtain $s_{\mathrm{Mi}}/m_{\mathrm{P}} \leq 0.01$ is $1/100$. Furthermore, the field s remains practically frozen during the inter-inflationary period since the Hubble parameter is larger than its mass.

- If s acquires effective mass of the order of H_{HI0} (as is [25, 26] generally expected) via the SUGRA scalar potential in Eq. (9), the field s can decrease to small values until the onset of MI. In our analysis we assume that:

- The inflaton S has minimal Kähler potential $K_m = |S|^2$ and therefore, induces [25] an effective mass to s during FHI, $m_s|_{HI} = \sqrt{3}H_{HI0}$.
- The modulus s is decoupled from the visible sector superfields both in Kähler potential and superpotential and has canonical Kähler potential, $K_s = s^2/2$. In such a simplified case, the value s_{min} at which the SUGRA potential has a minimum is [28] $s_{min} = 0$.

Following Refs. [34, 41], the evolution of s can be found by solving its e.o.m. More explicitly, inserting into Eq. (34),

- $H = H_{HI0}$ and $V = (m_s|_{HI})^2 s^2/2$ with $(m_s|_{HI})^2 = 3H_{HI0}^2$, we can derive the value of s at the end of FHI:

$$s_{Hf} = s_{Hi}e^{-3N_{HI}/2}\left(\cos\frac{\sqrt{3}}{2}N_{HI} + \sin\frac{\sqrt{3}}{2}N_{HI}\right), \qquad (51)$$

where $s_{Hi} \sim m_P$ is the value of s at the onset of FHI and N_{HI} is the total number of e-foldings obtained during FHI. We have also imposed the initial conditions, $s(N = 0) = s_{Hi}$ and $ds(N = 0)/dN = 0$.
- $H = H_{HI0}e^{-3\bar{N}/2}$ with $\bar{N} = \ln(R/R_{Hf})$ and $V = (m_s|_{MD})^2 s^2/2$ with $(m_s|_{MD})^2 = 3H^2/2$, we can derive the value of s at the beginig of MI:

$$s_{Mi} = s_{Hf}\left(\frac{V_{MI0}}{V_{HI0}}\right)^{1/4}\left(\cos\frac{\sqrt{15}}{12}\ln\frac{V_{HI0}}{V_{MI0}} + \sqrt{\frac{3}{5}}\sin\frac{\sqrt{15}}{12}\ln\frac{V_{HI0}}{V_{MI0}}\right), \quad (52)$$

where we have taken into account that during the inter-inflationary MD epoch $R \propto \rho^{-1/3}$ and imposed the initial conditions, $s(\bar{N} = 0) = s_{Hf}$ and $ds(\bar{N} = 0)/d\bar{N} = 0$.

In conclusion, combining Eqs. (51) and (52) we find

$$s_{Mi} \simeq m_P\left(\frac{V_{MI0}}{V_{HI0}}\right)^{1/4}e^{-3N_{HI}/2}. \qquad (53)$$

(vi)   In our analysis we have to ensure that the homogeneity of our present universe is not jeopardized by the quantum fluctuations of s during FHI which enter the horizon of MI, $\delta s|_{HMI}$ and during MI $\delta s|_{MI}$. Therefore, we have to dictate

$$s_{Mi} \gg \delta s|_{HMI} \quad \text{and} \quad s_{Mi} \gg \delta s|_{MI} \simeq H_s/2\pi. \qquad (54)$$

In order to estimate $\delta s|_{HMI}$, we find it convenient to single out the cases:

- If s does not acquire mass before MI, $\delta s|_{HMI}$ remains frozen during FHI and the inter-inflationary era. Consequently, we get

$$\delta s|_{HMI} \simeq H_{HI0}/2\pi. \qquad (55)$$

Obviously the first inequality in Eq. (54) is much more restrictive than the second one since $H_{HI0} \sim 10^{10}$ GeV whereas $H_s \sim m_s$.

- If s acquires mass before MI, we find [34, 41]:

$$\delta s|_{\text{HMI}} \simeq \frac{H_{\text{HI0}}}{2\pi} \left(\frac{H_{\text{HI0}}}{m_s|_{\text{HI}}}\right)^{1/2} e^{-3N_{\text{HIc}}/2} \left(\frac{V_{\text{MI0}}}{V_{\text{HI0}}}\right)^{1/4} = \frac{H_s}{3^{1/4}2\pi}, \qquad (56)$$

where Eq. (46) has been applied. As a consequence, the second inequality in Eq. (54) is roughly more restrictive than the first one and leads via Eq. (53) to the restriction:

$$N_{\text{HI}} \leq N_{\text{HI}}^{\max} \quad \text{with} \quad N_{\text{HI}}^{\max} = -\frac{2}{3} \ln \frac{(V_{\text{HI0}} V_{\text{MI0}})^{1/4}}{2\sqrt{3}\pi m_{\text{P}}^2}. \qquad (57)$$

Given that $V_{\text{HI0}}^{1/4} \sim 10^{14}$ GeV and $V_{\text{MI0}}^{1/4} \sim 10^{10}$ GeV, we expect $N_{\text{HI}}^{\max} \sim (15 - 18)$. This result signalizes an ugly tuning since it would be more reasonable FHI has a long duration due to the flatness of V_{HI}. This tuning could be evaded in a more elaborated set-up which would assure that $s_{\min} \neq 0$, due to the fact that s would not be completely decoupled – as in Refs. [34, 41].

(vii) If s decays exclusively through gravitational couplings, its decay width Γ_s and, consequently, T_{Mrh} are highly suppressed [42, 43]. In particular,

$$\Gamma_s = \frac{1}{8\pi} \frac{m_s^3}{m_{\text{P}}^2} \quad \text{and [45]} \quad T_{\text{Mrh}} = \left(\frac{72}{5g_{\rho*}(T_{\text{Mrh}})}\right)^{1/4} \sqrt{\Gamma_s m_{\text{P}}/\pi} \qquad (58)$$

with $g_{\rho*}(T_{\text{Mrh}}) \simeq 76$. For $m_s \sim 1$ TeV, we obtain $T_{\text{Mrh}} \simeq 10$ keV which spoils the success of NS within SBB, since RD era must have already begun before NS takes place at $T_{\text{NS}} \simeq 1$ MeV. This is [42] the well known moduli problem. The easiest (although somehow tuned) resolution to this problem is [42, 43] the imposition of the condition (for alternative proposals see Refs. [28, 43]):

$$m_s \geq 100 \, \text{TeV} \quad \text{which ensures} \quad T_{\text{Mrh}} \geq T_{\text{NS}}. \qquad (59)$$

To avoid the so-called [44] moduli-induced gravitino problem too, $m_{3/2}$ is to increase accordingly.

4.3. Numerical Results

In addition to the parameters mentioned in Sec. 2.6., our numerical analysis depends on the parameters:

$$m_{3/2}, \ m_s, \ m_s/H_s, \ s_{\text{Mi}}.$$

We take throughout $m_{3/2} = m_s = 100$ TeV which results to $T_{\text{Mrh}} = 1.5$ MeV through Eq. (58) and assures the satisfaction of the NS constraint with almost the lowest possible m_s. Since T_{Mrh} appears in Eq. (44) through its logarithm, its variation has a minor influence on the value of N_{tot} and, therefore, on our results. On the contrary, the hierarchy between $m_{3/2}$ and m_s plays an important role, because N_{MI} depends crucially only on F_s – see Eq. (35) – which in turn depends on the ratio m_s/H_s with $H_s \sim m_{3/2}$. As justified in the point (vii) we consider the choice $m_s \sim m_{3/2}$ as the most natural. It is worth mentioning, finally, that the chosen value of m_s (and $m_{3/2}$) has a key impact on the allowed parameter space of this scenario, when s does not acquire mass before MI. This is, because m_s is explicitly related to V_{MI0} – see Eq. (33) – which, in turn, is involved in Eq. (50) and constrains strongly H_{HI0} – see point (i) below.

As in Sec. 2.6., we use as input parameters κ (for standard and shifted FHI with fixed $M_S = 5 \times 10^{17}$ GeV) or M_S (for smooth FHI) and σ_*. Employing Eqs. (15) and (19), we can extract n_s and v_G respectively. For every chosen κ or M_S, we then restrict σ_* so as to achieve n_s in the range of Eq. (1) and take the output values of N_{HI*} (contrary to our strategy in Sec. 2.6. in which N_{HI*} given by Eq. (20) is treated as a constraint and n_s is an output parameter). Finally, for every given s_{Mi}, we find from Eq. (44) the required N_{MI} and the corresponding v_s or m_s/H_s from Eq. (41). Replacing F_s from Eqs. (35) in Eq. (41) and solving w.r.t m_s/H_s, we find:

$$\frac{m_s}{H_s} = \sqrt{\frac{1}{N_{MI}} \ln \frac{m_P}{s_{Mi}} \left(\frac{1}{N_{MI}} \ln \frac{m_P}{s_{Mi}} + 3 \right)} \qquad (60)$$

As regards the value of s_{Mi} we distinguish, once again, the cases:

(i) If s remains massless before MI, we choose $s_{Mi}/m_P = 0.01$. This value is close enough to m_P to have a non-negligible probability to be achieved by the randomization of s during FHI (see point (v) in Sec. 4.2.). At the same time, it is adequately smaller than m_P to guarantee good accuracy of Eqs. (35) and (41) near the interesting solutions and justify the fact that we neglect the uncertainty from the terms in the ellipsis in Eq. (32) – since we can obtain $N_{MI} \sim 30$ with low m_s/H_s's which assures low ϵ_s's as we emphasize in Eq. (38). Moreover, larger s_{Mi}'s lead to smaller parameter space for interesting solutions (with n_s near its central value).

Our results are presented in Figs. 13 – 16 for standard FHI (with N = 2) and in Table 4 for shifted and smooth FHI. Let us discuss each case separately:

- Standard FHI. We present the regions allowed by Eqs. (1), (19), (44), (47) – (50), (54) and (59) in the $\kappa - v_G$ (Fig. 13), $\kappa - m_s/H_s$ (Fig. 14), $\kappa - N_{HI*}$ (Fig. 15), and $\kappa - N_{MI}$ (Fig. 16) plane. The conventions adopted for the various lines are displayed in the r.h.s of every graph. In particular, the black solid [dashed] lines correspond to $n_s = 0.99$ [$n_s = 0.926$] whereas the gray solid lines have been obtained by fixing $n_s = 0.958$ – see Eq. (1). The dot-dashed [double dot-dashed] lines correspond to the lower bound on V_{HI0} [N_{MI}] from Eq. (50) [Eq. (49)]. The bold [faint] dotted lines correspond to $\alpha_s = -0.01$ [$\alpha_s = -0.005$]. Let us notice that:

  - The resulting v_G's and κ's are restricted to rather large values (although $v_G < M_{GUT}$) compared to those allowed within the other scenaria with one inflationary epoch (compare with Figs. 5 and 8). As a consequence, the SUGRA corrections in Eq. (10) play an important role.

  - The lower bound on V_{HI0} from Eq. (50) cut out sizeable slices of the allowed regions presented in Ref. [22]. This is due to the fact that we take here a much larger m_s in order to fulfill Eq. (59) – not considered in Ref. [22].

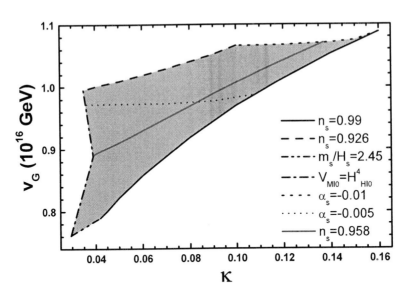

Figure 13. Allowed (lightly gray shaded) region in the $\kappa - v_G$ plane for standard FHI followed by MI realized by a field which remains massless before MI. The conventions adopted for the various lines are also shown.

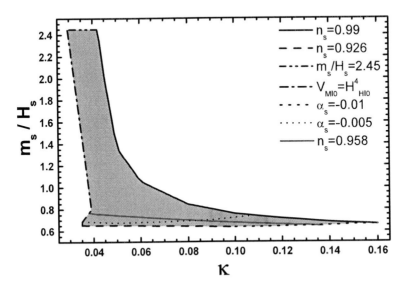

Figure 14. Allowed (lightly gray shaded) regions in the $\kappa - m_s/H_s$ plane for standard FHI followed by MI realized by a field which remains massless before MI. The conventions adopted for the various lines are also shown.

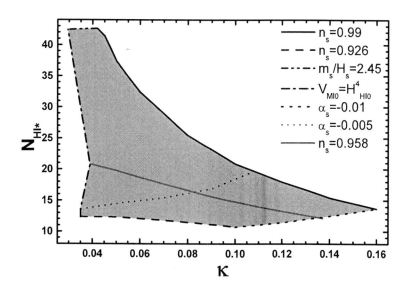

Figure 15. Allowed (lightly gray shaded) region in the $\kappa - N_{HI*}$ plane for standard FHI followed by MI realized by a field which remains massless before MI. The conventions adopted for the various lines are also shown.

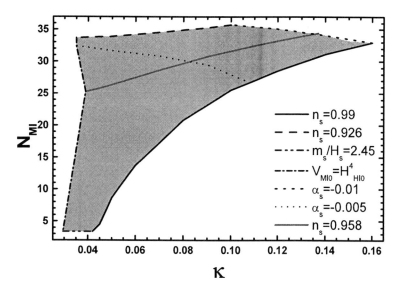

Figure 16. Allowed (lightly gray shaded) region in the $\kappa - N_{MI}$ plane for standard FHI followed by MI realized by a field which remains massless before MI. The conventions adopted for the various lines are also shown.

Table 4. Input and output parameters consistent with Eqs. (19), (44), (47) – (50), (54) and (59) in the cases of shifted ($M_{\mathrm{S}} = 5 \times 10^{17}$ GeV) or smooth FHI for $s_{\mathrm{Mi}}/m_{\mathrm{P}} = 0.01$, the nearest to M_{GUT} v_G and selected n_{s}'s within the mSUGRA double inflationary scenario when the inflaton of MI does not acquire effective mass.

SHIFTED FHI				SMOOTH FHI			
n_{s}	0.926	0.958	0.99	n_{s}	0.926	0.958	0.99
$v_G/10^{16}$ GeV	5.86	6.4	6.91	$v_G/10^{16}$ GeV	2.86	2.86	2.86
κ	0.035	0.04	0.045	$M_{\mathrm{S}}/5 \times 10^{17}$ GeV	0.815	0.87	0.912
$\sigma_*/10^{16}$ GeV	6.97	11.3	20.15	$\sigma_*/10^{16}$ GeV	22.18	23.53	25.54
$M/10^{16}$ GeV	4.57	4.92	5.24	$\mu_{\mathrm{S}}/10^{16}$ GeV	0.2	0.188	0.179
$1/\xi$	4.2	4.13	4.09	$\sigma_{\mathrm{f}}/10^{16}$ GeV	13.43	13.43	13.43
$N_{\mathrm{HI}*}$	12.75	20.8	40.45	$N_{\mathrm{HI}*}$	13.6	18	26
$-\alpha_{\mathrm{s}}/10^{-3}$	6	2.5	1	$-\alpha_{\mathrm{s}}/10^{-3}$	9	5.5	3
N_{MI}	31.1	23.1	3.35	N_{MI}	30.3	25.6	17.8
m_s/H_s	0.68	0.8	2.45	m_s/H_s	0.69	0.75	0.92

– The requirement of Eq. (47) does not constrain the parameters since it is overshadowed by the constraint of Eq. (50).

– In almost the half of the available parameter space for $n_{\mathrm{s}} \sim 0.958$ we have relatively high $|\alpha_{\mathrm{s}}|$, $0.005 \lesssim |\alpha_{\mathrm{s}}| \lesssim 0.01$.

– For $n_{\mathrm{s}} = 0.958$, we obtain $0.04 \lesssim \kappa \lesssim 0.14$, $0.89 \lesssim v_G/(10^{16}$ GeV$) \lesssim 1.08$ and $0.003 \lesssim |\alpha_{\mathrm{s}}| \lesssim 0.01$. Also, $12 \lesssim N_{\mathrm{HI}*} \lesssim 21.7$, $35 \gtrsim N_{\mathrm{MI}} \gtrsim 28$ and $0.64 \lesssim m_s/H_s \lesssim 0.74$. So, the interesting solutions correspond to slow rather than fast-roll MI.

• Shifted FHI. We list input and output parameters consistent with Eqs. (19), (44), (47) – (50), (54) and (59) for the nearest to M_{GUT} v_G and selected n_{s}'s in Table 4. The values of v_G come out considerably larger than in the case of standard FHI. However, the satisfaction of Eq. (50) in conjunction with Eq. (59) leads to $v_G > M_{\mathrm{GUT}}$. Indeed, $v_G = M_{\mathrm{GUT}}$ occurs for low κ's which produce V_{HI0}'s inconsistent with Eq. (50) – compare with Ref. [22].

• Smooth FHI. We arrange input and output parameters consistent with Eqs. (19), (44), (47) – (50), (54) and (59) for $v_G = M_{\mathrm{GUT}}$ and selected n_{s}'s in Table 4. In contrast with standard and shifted FHI, we can achieve $v_G = M_{\mathrm{GUT}}$ for every n_{s} in the range of Eq. (1). The mSUGRA corrections in Eq. (10) play an important role for every M_{S} encountered in Table 4 and $|\alpha_{\mathrm{s}}|$ is considerably enhanced but compatible with Eq. (48).

(ii)   If s acquires mass, s_{Mi} can be evaluated from Eq. (53). However, due to our ignorance of N_{HI}, there is an uncertainty in the determination of m_s/H_s, i.e. for every N_{MI}

required by Eq. (44), we can derive a maximal [minimal], $m_s/H_s|_{\max}$ [$m_s/H_s|_{\min}$], value of m_s/H_s. Eq. (60) implies that $m_s/H_s|_{\max}$ [$m_s/H_s|_{\min}$] is obtained by using the minimal [maximal] possible value of s_{Mi} which corresponds to $N_{\mathrm{HI}} = N_{\mathrm{HI}}^{\max}$ [$N_{\mathrm{HI}} = N_{\mathrm{HI}*}$]. Our results are presented in Figs. 17 – 20 for standard FHI (with $N = 2$) and in Table 5 for shifted and smooth FHI. Let us discuss each case separately:

- Standard FHI. We present the regions allowed by Eqs. (1), (19), (44) and (47) – (49), (57) and (59) in the $\kappa - v_G$ (Fig. 17), $\kappa - m_s/H_s$ (Fig. 18), $\kappa - N_{\mathrm{HI}*}$ (Fig. 19), and $\kappa - N_{\mathrm{MI}}$ (Fig. 20) plane. The conventions adopted for the various lines are displayed in the r.h.s of every graph. In particular, the black solid [dashed] lines correspond to $n_s = 0.99$ [$n_s = 0.926$] whereas the gray solid lines have been obtained by fixing $n_s = 0.958$ – see Eq. (1). The dot-dashed [double dot-dashed] lines correspond to the lower [upper] bound on $N_{\mathrm{HI}*}$ from Eq. (47) [Eq. (57)]. The double dot-dashed lines correspond to the upper [lower] bound on m_s/H_s [N_{MI}] from Eq. (36) [Eq. (49)]. The bold [faint] dotted lines correspond to $\alpha_s = -0.01$ [$\alpha_s = -0.005$]. Let us notice that:

  - Lower than those seen in Fig. 13 (but still larger than those shown in Figs. 5 and 8) v_G's and κ's are allowed in Fig. 17, since the constraint of Eq. (50) is not applied here. As κ increases above 0.01 the mSUGRA corrections in Eq. (10) become more and more significant.

  - The constraint from the upper bound on N_{HI} in Eq. (57) is very restrictive and almost overshadows this from the lower bound on N_{MI} in Eq. (49) (which is applied, e.g., only in the upper left corner of the allowed region in Fig. 18).

  - In contrast with the case (i), $0.005 \lesssim |\alpha_s| \lesssim 0.01$ holds only in a very limited part of the allowed regions.

  - For $n_s = 0.958$, we obtain $0.0035 \lesssim \kappa \lesssim 0.0085$ and $0.77 \lesssim v_G/(10^{16}\ \mathrm{GeV}) \lesssim 0.85$, or $0.08 \lesssim \kappa \lesssim 0.14$ and $0.96 \lesssim v_G/(10^{16}\ \mathrm{GeV}) \lesssim 1.08$. Also $0.002 \lesssim |\alpha_s| \lesssim 0.01$, $8.5 \lesssim N_{\mathrm{HI}*} \lesssim 17.3$, $34.3 \gtrsim N_{\mathrm{MI}} \gtrsim 26$ and $(1.4 - 1.96) \lesssim m_s/H_s \lesssim 2.35$. So, the interesting solutions correspond to fast rather than slow-roll MI.

- Shifted FHI. We list input and output parameters consistent with Eqs. (19), (44) and (47) – (49), (57) and (59) for the nearest to M_{GUT} v_G and selected n_s's in Table 5. The values of v_G come out again considerably larger than in the case of standard FHI. However, we take $v_G = M_{\mathrm{GUT}}$ only for $n_s = 0.926$ since the satisfaction of Eq. (57) requires $v_G < M_{\mathrm{GUT}}$ [$v_G > M_{\mathrm{GUT}}$] for $n_s = 0.958$ [$n_s = 0.99$]. The closest to M_{GUT} values of v_G for $n_s = 0.958$ and 0.99 are attained for $N_{\mathrm{HI}*} = N_{\mathrm{HI}}^{\max}$ and so, $m_s/H_s|_{\min} = m_s/H_s|_{\max}$.

- Smooth FHI. We display input and output parameters consistent with Eqs. (19), (44) and (47) – (49), (57) and (59) for the nearest to M_{GUT} v_G and selected n_s's in the Table 5. The results are quite similar to those for shifted FHI except for the fact that we have $v_G > M_{\mathrm{GUT}}$ for $n_s = 0.958$ and 0.99 and that $|\alpha_s|$ remains considerably enhanced.

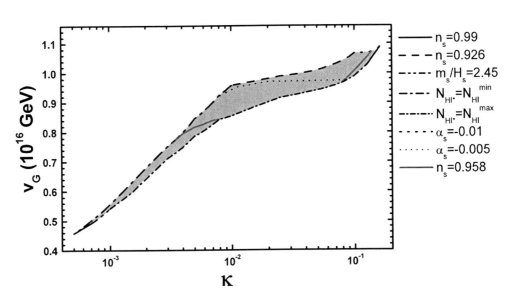

Figure 17. Allowed (lightly gray shaded) region in the $\kappa - v_G$ plane for standard FHI followed by MI realized by a field which acquires effective mass before MI. The conventions adopted for the various lines are also shown.

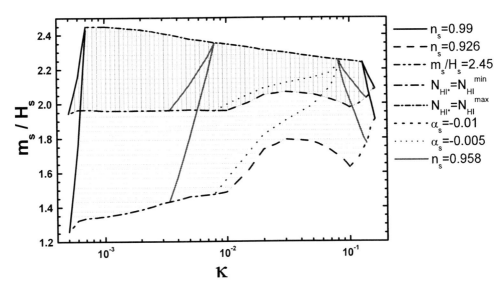

Figure 18. Allowed regions in the $\kappa - m_s/H_s$ plane for $N_{\mathrm{HI}} = N_{\mathrm{HI}}^{\mathrm{max}}$ (dark gray hatched region) or $N_{\mathrm{HI}} = N_{\mathrm{HI}*}$ (lightly gray ruled region) and standard FHI followed by MI realized by a field which acquires effective mass before MI. The conventions adopted for the various lines are also shown.

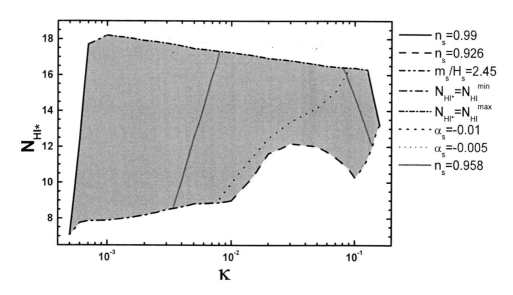

Figure 19. Allowed (lightly gray shaded) region in the $\kappa - N_{\mathrm{HI*}}$ plane for standard FHI followed by MI realized by a field which acquires effective mass before MI. The conventions adopted for the various lines are also shown.

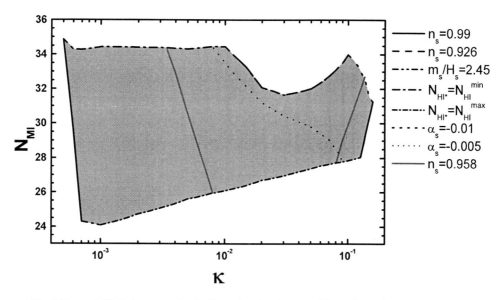

Figure 20. Allowed (lightly gray shaded) region in the $\kappa - N_{\mathrm{MI}}$ plane for standard FHI followed by MI realized by a field which acquires effective mass before MI. The conventions adopted for the various lines are also shown.

Table 5. Input and output parameters consistent with Eqs. (19), (44) and (47) – (49), (57) and (59) in the cases of shifted ($M_S = 5 \times 10^{17}$ GeV) or smooth FHI for the nearest to M_{GUT} v_G and selected n_s's within the mSUGRA double inflationary scenario when the inflaton of MI acquires effective mass before MI.

SHIFTED FHI				SMOOTH FHI			
n_s	0.926	0.958	0.99	n_s	0.926	0.958	0.99
$v_G/10^{16}$ GeV	2.86	1.93	12	$v_G/10^{16}$ GeV	2.86	3.3	4.61
κ	0.0106	0.0055	0.13	$M_S/5 \times 10^{17}$ GeV	0.815	1.06	1.66
$\sigma_*/10^{16}$ GeV	2.23	1.82	28.9	$\sigma_*/10^{16}$ GeV	22.18	25.73	32.8
$M/10^{16}$ GeV	2.38	1.65	8.95	$\mu_S/10^{16}$ GeV	0.2	0.21	0.25
$1/\xi$	4.67	5.04	4.05	$\sigma_f/10^{16}$ GeV	13.43	14.8	18.4
N_{HI*}	10.85	17.1	16.3	N_{HI*}	13.6	16.6	16.5
$-\alpha_s/10^{-3}$	5.6	1.9	7.3	$-\alpha_s/10^{-3}$	9	6.7	7.6
N_{MI}	32.6	26	28	N_{MI}	30.3	27.4	27.6
$m_s/H_s|_{max}$	2.03	2.3	2.45	$m_s/H_s|_{max}$	2.12	2.3	2.3
$m_s/H_s|_{min}$	1.6	2.3	2.45	$m_s/H_s|_{min}$	1.94	2.3	2.3

5. Conclusions

We reviewed the basic types (standard, shifted and smooth) of FHI in which the GUT breaking v.e.v, v_G, turns out to be comparable to SUSY GUT scale, M_{GUT}. Indeed, confronting these models with the restrictions on $P_{\mathcal{R}*}$ we obtain that v_G turns out a little lower than M_{GUT} for standard FHI whereas $v_G = M_{GUT}$ is possible for shifted and smooth FHI. However, the predicted n_s is just marginally consistent with the fitting of the WMAP3 data by the standard power-law ΛCDM cosmological model – if the horizon and flatness problems of SBB are resolved exclusively by FHI.

We showed that the results on n_s can be reconciled with data if we consider one of the following scenaria:

(i) FHI within qSUGRA. In this case, acceptable n_s's can be obtained by appropriately restricting the parameter c_q involved in the quasi-canonical Kähler potential, with a convenient sign. We paid special attention to the monotonicity of the inflationary potential which is crucial for the safe realization of FHI. Enforcing the monotonicity constraint, reduction of n_s below around 0.95 is prevented. Fixing in addition n_s to its central value, we found that (i) relatively large κ's but rather low v_G's are required within standard FHI with $0.013 \lesssim c_q \lesssim 0.03$ and (ii) $v_G = M_{GUT}$ is possible within smooth FHI with $c_q \simeq 0.0083$ but not within shifted FHI.

(ii) FHI followed by MI. In this case, acceptable n_s's can be obtained by appropriately restricting the number of e-foldings N_{HI*}. A residual number of e-foldings is produced by a bout of MI realized at an intermediate scale by a string modulus. We have taken into account extra restrictions on the parameters originating from:

- The resolution of the horizon and flatness problems of SBB.

- The requirements that FHI lasts long enough to generate the observed primordial fluctuations on all the cosmological scales and that these scales are not reprocessed by the subsequent MI.

- The limit on the running of n_s.

- The naturalness of MI.

- The homogeneity of the present universe.

- The complete randomization of the modulus if this remains massless before MI or its evolution before MI if it acquires effective mass.

- The establishment of RD before the onset of NS.

We discriminated two basic versions of this scenario, depending whether the modulus does or does not acquire effective mass before MI. We concluded that:

- If the modulus remains massless before MI, the combination of the randomization and NS constraints pushes the values of the inflationary plateau to relatively large values. Fixing n_s to its central value, we got (i) $v_G < M_{GUT}$ and $10 \lesssim N_{HI*} \lesssim 21.7$ within the standard FHI, (ii) $v_G > M_{GUT}$ and $N_{HI*} \simeq 21$ within shifted FHI and (iii) $v_G = M_{GUT}$ and $N_{HI*} \simeq 18$ within smooth FHI. In all cases, MI of the slow-roll type, with $m_s/H_s \sim (0.6-0.8)$, and a mild (of the order of 0.01) tuning of the initial value of the modulus produces the necessary additional number of e-foldings.

- If the modulus acquires effective mass before MI, lower values, than those encountered in the case (i), of the inflationary plateau are available. Fixing n_s to its central value, we got (i) $v_G < M_{GUT}$ and $8.5 \lesssim N_{HI*} \lesssim 17.5$ within the standard FHI and (ii) $v_G < M_{GUT}$ [$v_G > M_{GUT}$] and $N_{HI*} \simeq 17$ within shifted [smooth] FHI. In all cases, MI of the fast-roll type with $m_s/H_s \sim (1.4 - 2.45)$ and without any tuning of the initial value of the modulus produces the necessary additional number of e-foldings. However, FHI is constrained to be of short duration, producing a total number of e-foldings, $N_{HI} \lesssim 17$. This is rather questionable and can be evaded by introducing a more elaborated structure for the Kähler potential or superpotential of the modulus (see, e.g., Ref. [34, 41]).

Trying to compare the proposed methods for the reduction of n_s within FHI, we can do the following comments:

- The main advantage of the method in the case (i) is that the standard one-step inflationary cosmological set-up remains intact. This method becomes rather attractive when the minimum-maximum structure of the inflationary potential is avoided. However, the possible in this way decrease of n_s is rather limited.

- The method of the case (ii) offers a comfortable reduction of n_s but it requires a more complicate cosmological set-up with advantages (dilution of gravitinos and defects) and disadvantages (complications with baryogenesis). The most natural and simple version of this scenario is realized when the modulus remains massless during FHI since it requires a very mild tuning.

Hopefully, the proposed scenaria will be further probed by the measurements of the Planck satellite which is expected to give results on n_s with an accuracy $\Delta n_s \simeq 0.01$ by the end of the decade [46].

Acknowledgments

We would like to thank G. Lazarides and A. Pilaftsis for fruitful and pleasant collaborations, from which parts of this work are culled. This work was supported from the PPARC research grant PP/C504286/1.

References

[1] A.H. Guth, *Phys. Rev. D* **23**, 347 (1981).

[2] D.H. Lyth and A. Riotto, *Phys. Rept.* **314**, 1 (1999) [hep-ph/9807278].

[3] G. Lazarides, *Lect. Notes Phys.* **592**, 351 (2002) [hep-ph/0111328];
    G. Lazarides, *J. Phys. Conf. Ser.* **53**, 528 (2006) [hep-ph/0607032].

[4] A. Riotto, hep-ph/0210162.

[5] G. Lazarides, hep-ph/0011130.

[6] E.J. Copeland *et al.*, *Phys. Rev. D* **49**, 6410 (1994) [astro-ph/9401011].

[7] G.R. Dvali, Q. Shafi and R.K. Schaefer, *Phys. Rev. Lett.* **73**, 1886 (1994) [hep-ph/9406319];
    G. Lazarides, R.K. Schaefer, and Q. Shafi, *Phys. Rev. D* **56**, 1324 (1997) [hep-ph/9608256].

[8] R. Jeannerot *et al.*, *J. High Energy Phys.* **10**, 012 (2000) [hep-ph/0002151].

[9] G. Lazarides and C. Panagiotakopoulos, *Phys. Rev. D* **52**, 559 (1995) [hep-ph/9506325];
    G. Lazarides *et al.*, *Phys. Rev. D* **54**, 1369 (1996) [hep-ph/9606297];
    R. Jeannerot, S. Khalil, and G. Lazarides, *Phys. Lett. B* **506**, 344 (2001) [hep-ph/0103229].

[10] A.D. Linde and A. Riotto *Phys. Rev. D* **56**, 1841 (1997) [hep-ph/9703209];
     V.N. Şenoğuz and Q. Shafi, *Phys. Lett. B* **567**, 79 (2003) [hep-ph/0305089].

[11] D.N. Spergel *et al.*, astro-ph/0603449.

[12] R.A. Battye *et al.*, *J. Cosmol. Astropart. Phys.* **09**, 007 (2006) [astro-ph/0607339]

[13] G. Lazarides *et al.*, *Phys. Rev. D* **70**, 123527 (2005) [hep-ph/0409335].

[14] R. Jeannerot and M. Postma, *J. High Energy Phys.* **05**, 071 (2005) [hep-ph/0503146].

[15] J. Rocher and M. Sakellariadou, *J. Cosmol. Astropart. Phys.* **03**, 004 (2005) [hep-ph/0406120].

[16] B. Garbrecht *et al.*, *J. High Energy Phys.* **12**, 038 (2006) [hep-ph/0605264].

[17] M. Bastero-Gil *et al.*, *Phys. Lett. B* **651**, 345 (2007) [hep-ph/0604198].

[18] C. Panagiotakopoulos, *Phys. Lett. B* **402**, 257 (1997) [hep-ph/9703443].

[19] L. Boubekeur and D. Lyth, *J. Cosmol. Astropart. Phys.* **07**, 010 (2005) [hep-ph/0502047].

[20] M. ur Rehman, V.N. Şenoğuz, and Q. Shafi, *Phys. Rev. D* **75**, 043522 (2007) [hep-ph/0612023].

[21] G. Lazarides and A. Vamvasakis, arXiv:0705.3786.

[22] G. Lazarides and C. Pallis, *Phys. Lett. B* **651**, 216 (2006) [hep-ph/0702260];
G. Lazarides, arXiv:0706.1436.

[23] P. Binétruy and M.K. Gaillard, *Phys. Rev. D* **34**, 3069 (1986);
F.C. Adams *et al.*, *Phys. Rev. D* **47**, 426 (1993) [hep-ph/9207245];
T. Banks *et al.*, *Phys. Rev. D* **52**, 3548 (1995) [hep-ph/9503114];
R. Brustein *et al.*, *Phys. Rev. D* **68**, 023517 (2003) [hep-ph/0205042].

[24] G. Lazarides and A. Vamvasakis, arXiv:0709.3362

[25] M. Dine, L. Randall and S. Thomas, *Phys. Rev. Lett.* **75**, 398 (1995) [hep-ph/9503303];
M.K. Gaillard *et al.*, *Phys. Lett. B* **355**, 71 (1995) [hep-ph/9504307].

[26] D.H. Lyth and T. Moroi, *J. High Energy Phys.* **05**, 004 (2004) [hep-ph/0402174].

[27] A. Linde, *J. High Energy Phys.* **11**, 052 (2001) [hep-ph/0110195].

[28] G. Dvali, hep-ph/9503259.

[29] M.Yu. Khlopov and A.D. Linde, Phys. Lett. B **138**, 265 (1984);
J. Ellis, J.E. Kim, and D.V. Nanopoulos, *ibid.* **145**, 181 (1984).

[30] T.W.B. Kibble, *J. Phys. A* **9**, 387 (1976).

[31] G. Ballesteros *et al.*, *J. Cosmol. Astropart. Phys.* **03**, 001 (2006) [hep-ph/0601134].

[32] V.N. Şenoğuz and Q. Shafi, *Phys. Rev. D* **71**, 043514 (2005) [hep-ph/0412102].

[33] D.H. Lyth and D. Wands, *Phys. Lett. B* **524**, 5 (2002) [hep-ph/0110002];
T. Moroi and T. Takahashi, *Phys. Lett. B* **522** 215 (2001);
T. Moroi and T. Takahashi, *Phys. Lett. B* **539** 303(E) (2002) [hep-ph/0110096];
D. H. Lyth, C. Ungarelli and D. Wands, *Phys. Rev. D* **67** 023503 (2003) [astro-ph/0208055].

[34] M. Kawasaki *et al.*, *Phys. Rev. D* **68**, 023508 (2003) [hep-ph/0304161];
M. Yamaguchi and J. Yokoyama, *Phys. Rev. D* **68**, 123520 (2003) [hep-ph/0307373];
M. Yamaguchi and J. Yokoyama, *Phys. Rev. D* **70**, 023513 (2004) [hep-ph/0402282];
M. Kawasaki *et al.*, *Phys. Rev. D* **74**, 043525 (2006) [hep-ph/0605271].

[35] J. Garcia-Bellido *et al.*, *Phys. Rev. D* **60**, 123504 (1999) [hep-ph/9902449];
L.M. Krauss and M. Trodden, *Phys. Rev. Lett.* **83**, 1502 (1999) [hep-ph/9902420].

[36] K. Benakli and S. Davidson, *Phys. Rev. D* **60**, 025004 (1999) [hep-ph/9810280].

[37] G. Dvali and S. Kachru, hep-ph/0310244.

[38] C.P. Burgess *et al.*, *J. High Energy Phys.* **05**, 067 (2005) [hep-th/0501125].

[39] U. Seljak, *et al.*, *J. Cosmol. Astropart. Phys.* **10**, 014 (2006) [astro-ph/0604335].

[40] A.A. Starobinsky and J. Yokoyama, *Phys. Rev. D* **50**, 6357 (1994);
E.J. Chun *et al.*, *Phys. Rev. D* **70**, 103510 (2004) [hep-ph/0402059].

[41] K.I. Izawa *et al.*, *Phys. Lett. B* **411**, 249 (1997) [hep-ph/9707201];
M. Kawasaki *et al.*, *Phys. Rev. D* **57**, 6050 (1998) [hep-ph/9710259];
M. Kawasaki and T. Yanagida, *Phys. Rev. D* **59**, 043512 (1999) [hep-ph/9807544].

[42] G.D. Coughlan *et al.*, *Phys. Lett. B* **131** (1983) 59;
B. de Carlos *et al.*, *Phys. Lett. B* **318**, 447 (1993) [hep-ph/9308325].

[43] M. Berkooz *et al.*, *Phys. Rev. D* **71** 103502 (2005) [hep-ph/0409226].

[44] M. Endo *et al.*, *Phys. Rev. Lett.* **96**, 211301 (2006) [hep-ph/0602061];
S. Nakamura and M. Yamaguchi, *Phys. Lett. B* **638**, 389 (2006) [hep-ph/0602081].

[45] C. Pallis, *Nucl. Phys.* **B751**, 129 (2006) [hep-ph/0510234].

[46] http://www.rssd.esa.int/SA/PLANCK/include/report/redbook/redbook-science.htm

In: High Energy Physics Research Advances
Editors: T.P. Harrison et al, pp. 39-79
ISBN 978-1-60456-304-7
© 2008 Nova Science Publishers, Inc.

Chapter 2

GENERALIZED SINE-GORDON MODEL AND BARYONS IN TWO-DIMENSIONAL QCD

Harold Blas
Departamento de Física.
Univerdidade Federal de Mato Grosso (UFMT)
Av. Fernando Correa, s/n, Coxipó, Cuiabá
78060-900, Mato Grosso-Brazil

Abstract

We consider the sl(3,C) affine Toda model coupled to matter (Dirac spinor) (ATM) and through a gauge fixing procedure we obtain the classical version of the generalized sl(3,C) sine-Gordon model (cGSG) which completely decouples from the Dirac spinors. The GSG models are multifield extensions of the ordinary sine-Gordon model. In the spinor sector we are left with Dirac fields coupled to cGSG fields. Based on the equivalence between the U(1) vector and topological currents, which holds in the theory, it is shown the confinement of the spinors inside the solitons and kinks of the cGSG model providing an extended hadron model for "quark" confinement [JHEP0701(2007)027]. Moreover, the solitons and kinks of the generalized sine-Gordon (GSG) model are shown to describe the normal and exotic baryon spectrum of two-dimensional QCD. The GSG model arises in the low-energy effective action of bosonized QCD2 with unequal quark mass parameters [JHEP0703(2007)055]. The GSG potential for three flavors resembles the potential of the effective chiral lagrangian proposed by Witten to describe low-energy behavior of four dimensional QCD. Among the attractive features of the GSG model are the variety of soliton and kink type solutions for QCD2 unequal quark mass parameters. Exotic baryons in QCD2 [Ellis and Frishman, JHEP0508(2005)081] are discussed in the context of the GSG model. Various semi-classical computations are performed improving previous results and clarifying the role of unequal quark masses. The remarkable double sine Gordon model also arises as a reduced GSG model bearing a kink(K) type solution describing a multi-baryon.

1. Introduction

The sine-Gordon model (SG) has been studied over the decades due to its many properties and mathematical structures such as integrability and soliton solutions. It can be used as a toy model for non-perturbative quantum field theory phenomena. In this context, some extensions and modifications of the SG model deserve attention. An extension taking multi-frequency terms as the potential has been investigated in connection to various physical applications [1, 2, 3, 4].

Besides, an extension defined for multi-fields is the so-called generalized sine-Gordon model (GSG) which has been found in the study of the strong/weak coupling sectors of the $sl(N,\mathbb{C})$ affine Toda model coupled to matter fields (ATM) [5, 6]. In connection to these developments, the bosonization process of the multi-flavor massive Thirring model (GMT) provides the quantum version of the (GSG) model [7]. The GSG model provides a framework to obtain (multi-)soliton solutions for unequal mass parameters of the fermions in the GMT sector and study the spectrum and their interactions. The extension of this picture to the NC space-time has been addressed (see [8] and references therein).

In the first part of this chapter we study the spectrum of solitons and kinks of the GSG model proposed in [5, 6, 7] and consider the closely related ATM model from which one gets the classical GSG model (cGSG) through a gauge fixing procedure [9]. Some reductions of the GSG model to one-field theory lead to the usual SG model and to the so-called multi-frequency sine-Gordon models. In particular, the double (two-frequency) sine-Gordon model (DSG) appears in a reduction of the $sl(3,\mathbb{C})$ GSG model. The DSG theory is a nonintegrable quantum field theory with many physical applications [3, 4].

Once a convenient gauge fixing is performed by setting to constant some spinor bilinears in the ATM model we are left with two sectors: the cGSG model which completely decouples from the spinors and a system of Dirac spinors coupled to the cGSG fields. In refs. [10, 11] the authors have proposed a $1 + 1$-dimensional bag model for quark confinement, here we follow their ideas and generalize for multi-flavor Dirac spinors coupled to cGSG solitons and kinks. The first reference considers a model similar to the $sl(2)$ ATM theory, and in the second one the DSG kink is proposed as an extended hadron model.

Recently, in QCD$_4$ there appeared some puzzles related with unequal quark masses [12] providing an extra motivation to consider QCD$_2$ as a testing ground for non-perturbative methods that might have relevance in the real world. Claims for the existence of exotic baryons - that can not be composed of just three quarks - have inspired intense studies of the theory and phenomenology of QCD in the strong-interaction regime. In particular, it has led to the discovery that the strong coupling regime may contain unexpected correlations among groups of two or three quarks and antiquarks. Results of growing number of experiments at laboratories around the world provide contradictive situation regarding the experimental observation of possible pentaquark states, see e.g. [13]. These experiments have thus opened new lines of theoretical investigation that may survive even if the original inspiration - the exotic Θ^+ pentaquark existence- is not confirmed. After the reports of null results started to accumulate the initial optimism declined, and the experimental situation remains ambiguous to the present. The increase in statistics led to some recent new claims for positive evidence [14], while the null result [15] by CLAS is specially significant because it contradicts their earlier positive result, suggesting that at least in their case the original claim was an artifact due to low statistics. All this experimental activity spurred a great amount of theoretical work in all kinds of models for hadrons and a renewed interest in soliton models. Recently, there is new strong evidence of an extremely narrow Θ^+ resonance from DIANA collaboration and a very significant new evidence from LEPS. For a recent account of the theoretical and experimental situation see e.g. [16].

On the other hand, Quantum Chromodynamics in two-dimensions (QCD $_2$) (see e.g. [17]) has long been considered a useful theoretical laboratory for understanding non-perturbative strong-interaction problems such as confinement [18], the large-N $_c$ expansion [19], baryon structure [20] and, more recently, the chiral-soliton picture for normal and exotic baryons [21, 22]. Even though there are various differences between QCD $_4$ and QCD$_2$, this theory may provide interesting insights into the physical four-dimensional world. In two dimensions, an exact and complete bosonic description exists and in the strong-coupling limit one can eliminate the color degrees of freedom entirely, thus getting an effective action expressed in terms of flavor degrees of freedom only. In this way various aspects have been studied, such as baryon spectrum and its $\bar{q}q$ content [20]. The constituent quark solitons of baryons were uncovered taking into account the both bosonized flavor and color degrees of freedom [23]. In particular, the study of meson-baryon scattering and resonances is a nontrivial task for unequal quark masses even in 2D [24].

It has been conjectured that the low-energy action of QCD$_2$ ($e_c >> M_q$, M_q quark mass and e_c gauge coupling) might be related to massive two dimensional integrable models, thus leading to the exact solution of the strong coupled QCD $_2$ [20]. As an example of this picture, it has been shown that the so-called $su(2)$ affine Toda model coupled to matter (Dirac) field (ATM) [25] describes the low-energy spectrum of QCD$_2$ (one flavor and N_C colors) [26]. The ATM model allowed the exact computation of the string tension in QCD $_2$ [26], improving the approximate result of [27]. The strong coupling sector of the $su(2)$ ATM model is described by the usual sine-Gordon model [30, 28, 29]. The baryons in QCD may be described as solitons in the bosonized formulation. In the strong-coupling limit the static classical soliton which describes a baryon in QCD$_2$ turns out to be the ordinary sine-Gordon soliton, i.e.

$$\Phi(x) = \frac{4}{\beta_0}\tan^{-1}[\exp \beta_0 \sqrt{2}\widetilde{m}x] \tag{1}$$

where $\beta_0 = \sqrt{\frac{4\pi}{N_C}}$ is the coupling constant of the sine-Gordon theory, $8\sqrt{2}\widetilde{m}/\beta_0$ is the mass of the soliton, and $\widetilde{m}$ is related to the common bare mass of the quarks by a renormalization group relation relevant to two dimensions. The soliton in (1) has non-zero baryon number as well as Y charge. The quantum correction to the soliton mass, obtained by time-dependent rotation in flavor space, is suppressed by a factor of N_C compared to the classical contribution to the baryon mass [20]. The considerations of more complicated mass matrices and higher order corrections to the $M_q/e_c \to 0$ limit are among the issues that deserve further attention.

In the second part of this chapter we show that various aspects of the low-energy effective QCD$_2$ action with unequal quark masses can be described by the generalized sine-Gordon model. The GSG model has appeared in the study of the strong coupling sector of the $sl(n,\mathbb{C})$ ATM theory[5, 6, 9], and in the bosonized multiflavor massive Thirring model [7]. In particular, the GSG model provides the framework to obtain (multi-)soliton solutions for unequal quark mass parameters. Choosing the normalization such that quarks have baryon number $Q_B^0 = 1$ and a one-soliton has baryon number N_C, we classify the configurations in the GSG model with baryon numbers N_C, $2N_C$, ... $4N_C$. For example, the double sine-Gordon model provides a kink type solution describing a multi-baryon state with baryon number $4N_C$. Then, using the GSG model we generalize the results of refs. [20, 21] which applied the semi-classical quantization method in order to uncover the normal [20] and exotic baryon [21] spectrum of QCD$_2$. One of the main features of the GSG model is that the one-soliton solution requires the QCD quark mass parameters to satisfy certain relationship. In two dimensions there are no spin degrees of freedom, so, the lowest-lying baryons are related to the purely symmetric Young tableau, the **10** dimensional representation of flavor $SU(3)$. This is the analogue of the multiplet containing the baryons Δ, Σ, Ξ, Ω^- in QCD$_4$. The next state corresponds to a state with the quantum numbers of four quarks and an antiquark, the so-called pentaquark, which in two dimensions forms a **35** representation of flavor $SU(3)$. This corresponds to the four dimensional multiplet $\overline{\mathbf{10}}$, which contain the exotic baryons Θ^+, $\bar{\Sigma}$, Ξ^{--}.

Here we improve the results of refs. [20, 21], such as the normal and exotic baryon masses, the relevant mass ratios and the radius parameter of the exotic baryons. The semi-classical computations of the masses get quantum corrections due to the unequal mass term contributions and to the form of the diagonal ansatz taken for the flavor field (related to GSG model) describing the lowest-energy state of the effective action. The corrections to the normal baryon masses are an increase of 3.5% to the earlier value obtained in [21], and in the case of the exotic baryon our computations improve the behavior of the quantum correction by decreasing the earlier value in 0.34 units, so making the semi-classical result more reliable. Let us mention that for the first exotic baryon [21] the quantum correction was greater than the classical term by a factor of 2.46, so that semi-classical approximation may not be a good approximation. As a curiosity, with the relevant values obtained by us for QCD$_2$ we computed the ratio between the lowest exotic baryon and the $R = \mathbf{10}$ baryon masses $M_{35}/M_{10} \sim 1.65$, which is only 1% larger than the analogous four dimensional QCD ratio $M_{\Theta^+}/M_{nucleon} \sim 1.63$. In [21] the relevant QCD$_2$ ratio was 17% larger than this value. The mass formulae for the normal and exotic baryons corresponding, respectively, to the representations **10** and **35**, in two dimensions resemble the general chiral-soliton model formula in four dimensions [31] except that there is no spin-dependent term $\sim J(J+1)$, and an analog term containing the soliton moment of inertia emerges.

In the next section we define the $sl(3,\mathbb{C})$ classical GSG model and describe some of its symmetries. In section 3. we consider the $sl(3,\mathbb{C})$ affine Toda model coupled to matter and obtain the cGSG model through a gauge fixing procedure. We discuss the physical properties of its spectrum. The topological charges are introduced, as well as the idea of baryons as solitons (or kinks), and the quark confinement mechanism is discussed. In section 4. it is summarized the bosonized low-energy effective action of QCD$_2$ and introduced the lowest-energy state described by the GSG action. The global QCD$_2$ symmetries are discussed. Section 5. provides the GSG solitons and kink solutions relevant to our QCD$_2$ discussions. In section 6. the semi-classical method of quantization relevant to a general diagonal ansatz is introduced. In subsection 6.1. we briefly review the ordinary sine-Gordon soliton semi-classical quantization in the context of QCD$_2$. In section 7. we discuss the quantum correction to the $SU(3)$ GSG ansatz in the framework of semi-classical quantization. In subsection 7.1. the GSG one-soliton state is rotated in $SU(3)$ flavor space by a time-dependent $A(t)$. In subsection 7.2. the lowest-energy baryon state with baryon number N_C is introduced. The possible vibrational modes are briefly discussed in subsection 7.3.. In section 8. it is discussed the first and higher multiplet exotic baryons and provided the relevant quantum corrections the their masses, the ratio M_{35}/M_{10}, and an estimate for the exotic baryon radius parameter. The last section presents a summary and discussions.

2. The Generalized Sine-Gordon Model (GSG)

The generalized sine-Gordon model (GSG) related to $sl(N,\mathbb{C})$ is defined by [5, 6, 7]

$$S = \int d^2x \sum_{i=1}^{N_f} \left[\frac{1}{2}(\partial_\mu \Phi_i)^2 + \mu_i \left(\cos\beta_i\Phi_i - 1 \right) \right]. \tag{2}$$

The Φ_i fields in (2) satisfy the constraints

$$\Phi_p = \sum_{i=1}^{N-1} \sigma_{pi}\Phi_i, \quad p = N, N+1, ..., N_f, \quad N_f = \frac{N(N-1)}{2}, \tag{3}$$

where σ_{pi} are some constant parameters and N_f is the number of positive roots of the Lie algebra $sl(N,\mathbb{C})$. In the context of the Lie algebraic construction of the GSG system these constraints arise from the relationship between the positive and simple roots of $sl(N,\mathbb{C})$. Thus, in (2) we have $(N-1)$ independent fields.

We will consider the $sl(3,\mathbb{C})$ case with two independent real fields $\varphi_{1,2}$, such that

$$\Phi_1 = 2\varphi_1 - \varphi_2; \quad \Phi_2 = 2\varphi_2 - \varphi_1; \quad \Phi_3 = r\,\varphi_1 + s\,\varphi_2, \quad s, r \in \mathbb{R} \tag{4}$$

which must satisfy the constraint

$$\beta_3\Phi_3 = \delta_1\beta_1\Phi_1 + \delta_2\beta_2\Phi_2, \quad \beta_i \equiv \beta_0\nu_i, \tag{5}$$

where β_0, ν_i, δ_1, δ_2 are some real numbers. Therefore, the $sl(3,\mathbb{C})$ GSG model can be regarded as three usual sine-Gordon models coupled through the linear constraint (5).

Taking into account (4)-(5) and the fact that the fields φ_1 and φ_2 are independent we may get the relationships

$$\nu_2\delta_2 = \rho_0\nu_1\delta_1 \quad \nu_3 = \frac{1}{r+s}(\nu_1\delta_1 + \nu_2\delta_2); \quad \rho_0 \equiv \frac{2s+r}{2r+s} \tag{6}$$

The $sl(3,\mathbb{C})$ model has a potential density

$$V[\varphi_i] = \sum_{i=1}^{3} \mu_i \left(1 - \cos\beta_i\Phi_i \right) \tag{7}$$

The GSG model has been found in the process of bosonization of the generalized massive Thirring model (GMT) [7]. The GMT model is a multiflavor extension of the usual massive Thirring model incorporating massive fermions with current-current interactions between them. In the $sl(3,\mathbb{C})$ construction of [7] the parameters δ_i depend on the couplings β_i and they satisfy certain relationship. This is obtained by assuming $\mu_i > 0$ and the zero of the potential given for $\Phi_i = \frac{2\pi}{\beta_i}n_i$, which substituted into (5) provides

$$n_1\delta_1 + n_2\delta_2 = n_3, \quad n_i \in \mathbb{Z} \tag{8}$$

The last relation combined with (6) gives

$$(2r+s)\frac{n_1}{\nu_1}+(2s+r)\frac{n_2}{\nu_2}=3\frac{n_3}{\nu_3}. \tag{9}$$

The periodicity of the potential implies an infinitely degenerate ground state and then the theory supports topologically charged excitations. The vacuum configuration is related to the fundamental weights [9]. For a future purpose let us consider the fields Φ_1 and Φ_2 and the vacuum lattice defined by

$$(\Phi_1,\ \Phi_2)=\frac{2\pi}{\beta_0}(\frac{n_1}{\nu_1},\ \frac{n_2}{\nu_2}),\quad n_a\in\mathbb{Z}. \tag{10}$$

It is convenient to write the equations of motion in terms of the independent fields φ_1 and φ_2

$$\begin{aligned}
\partial^2\varphi_1 &= -\mu_1\beta_1\Delta_{11}\sin[\beta_1(2\varphi_1-\varphi_2)]-\mu_2\beta_2\Delta_{12}\sin[\beta_2(2\varphi_2-\varphi_1)]+ \\
&\quad \mu_3\beta_3\Delta_{13}\sin[\beta_3(r\varphi_1+s\varphi_2)]
\end{aligned} \tag{11}$$

$$\begin{aligned}
\partial^2\varphi_2 &= -\mu_1\beta_1\Delta_{21}\sin[\beta_1(2\varphi_1-\varphi_2)]-\mu_2\beta_2\Delta_{22}\sin[\beta_2(2\varphi_2-\varphi_1)]+ \\
&\quad \mu_3\beta_3\Delta_{23}\sin[\beta_3(r\varphi_1+s\varphi_2)],
\end{aligned} \tag{12}$$

where

$$\begin{aligned}
A &= \beta_0^2\nu_1^2(4+\delta^2+\delta_1^2\rho_1^2r^2),\quad B=\beta_0^2\nu_1^2(1+4\delta^2+\delta_1^2\rho_1^2s^2), \\
C &= \beta_0^2\nu_1^2(2+2\delta^2+\delta_1^2\rho_1^2r\,s), \\
\Delta_{11} &= (C-2B)/\Delta,\quad \Delta_{12}=(B-2C)/\Delta,\quad \Delta_{13}=(r\,B+s\,C)/\Delta, \\
\Delta_{21} &= (A-2C)/\Delta,\quad \Delta_{22}=(C-2A)/\Delta,\quad \Delta_{23}=(r\,C+s\,A)/\Delta \\
\Delta &= C^2-AB,\quad \delta=\frac{\delta_1}{\delta_2}\rho_0,\quad \rho_1=\frac{3}{2r+s}
\end{aligned}$$

Notice that the eqs. of motion (11)-(12) exhibit the symmetries

$$\varphi_1\leftrightarrow\varphi_2,\quad \mu_1\leftrightarrow\mu_2,\quad \nu_1\leftrightarrow\nu_2,\quad \delta_1\leftrightarrow\delta_2,\quad r\leftrightarrow s \tag{13}$$

$$\varphi_a\leftrightarrow-\varphi_a,\quad a=1,2 \tag{14}$$

Some type of coupled sine-Gordon models have been considered in connection to various interesting physical problems [32]. For example a system of two coupled SG models has been proposed in order to describe the dynamics of soliton excitations in deoxyribonucleic acid (DNA) double helices [33]. In general these type of equations have been solved by perturbation methods around decoupled sine-Gordon exact solitons. In section 5. the system (11)-(12) will be shown to possess exact soliton and kink type solutions.

3.   Classical GSG as a Reduced Toda Model Coupled to Matter

In this section we obtain the classical GSG model as a reduced model starting from the $sl(3,\mathbb{C})$ affine Toda model coupled to matter fields (ATM) and closely follows ref. [9].

The previous treatments of the $sl(3,\mathbb{C})$ ATM model used the symplectic and on-shell decoupling methods to unravel the classical GSG and generalized massive Thirring (GMT) dual theories describing the strong/weak coupling sectors of the ATM model [5, 6, 29]. The ATM model describes some scalars coupled to spinor (Dirac) fields in which the system of equations of motion has a local gauge symmetry. In this way one includes the spinor sector in the discussion and conveniently gauge fixing the local symmetry by setting some spinor bilinears to constants we are able to decouple the scalar (Toda) fields from the spinors, the final result is a direct construction of the classical generalized sine-Gordon model (cGSG) involving only the scalar fields. In the spinor sector we are left with a system of equations in which the Dirac fields couple to the cGSG fields.

The conformal version of the ATM model is defined by the following equations of motion [25]

$$\frac{\partial^2 \theta_a}{4i\, e^\eta} = m_\psi^1[e^{\eta - i\phi_a}\widetilde{\psi}_R^l \psi_L^l + e^{i\phi_a}\widetilde{\psi}_L^l \psi_R^l] + m_\psi^3[e^{-i\phi_3}\widetilde{\psi}_R^3 \psi_L^3 + e^{\eta + i\phi_3}\widetilde{\psi}_L^3 \psi_R^3];$$
$$a = 1, 2 \tag{15}$$

$$-\frac{\partial^2 \widetilde{\nu}}{4} = im_\psi^1 e^{2\eta - \phi_1}\widetilde{\psi}_R^1 \psi_L^1 + im_\psi^2 e^{2\eta - \phi_2}\widetilde{\psi}_R^2 \psi_L^2 + im_\psi^3 e^{\eta - \phi_3}\widetilde{\psi}_R^3 \psi_L^3 + \mathbf{m}^2 e^{3\eta} \tag{16}$$

$$-2\partial_+ \psi_L^1 = m_\psi^1 e^{\eta + i\phi_1}\psi_R^1, \qquad -2\partial_+ \psi_L^2 = m_\psi^2 e^{\eta + i\phi_2}\psi_R^2, \tag{17}$$

$$2\partial_- \psi_R^1 = m_\psi^1 e^{2\eta - i\phi_1}\psi_L^1 + 2i\left(\frac{m_\psi^2 m_\psi^3}{im_\psi^1}\right)^{1/2} e^\eta(-\psi_R^3 \widetilde{\psi}_L^2 e^{i\phi_2} - \widetilde{\psi}_R^2 \psi_L^3 e^{-i\phi_3}), \tag{18}$$

$$2\partial_- \psi_R^2 = m_\psi^2 e^{2\eta - i\phi_2}\psi_L^2 + 2i\left(\frac{m_\psi^1 m_\psi^3}{im_\psi^2}\right)^{1/2} e^\eta(\psi_R^3 \widetilde{\psi}_L^1 e^{i\phi_1} + \widetilde{\psi}_R^1 \psi_L^3 e^{-i\phi_3}), \tag{19}$$

$$-2\partial_+ \psi_L^3 = m_\psi^3 e^{2\eta + i\phi_3}\psi_R^3 + 2i\left(\frac{m_\psi^1 m_\psi^2}{im_\psi^3}\right)^{1/2} e^\eta(-\psi_L^1 \psi_R^2 e^{i\phi_2} + \psi_L^2 \psi_R^1 e^{i\phi_1}), \tag{20}$$

$$2\partial_- \psi_R^3 = m_\psi^3 e^{\eta - i\phi_3}\psi_L^3, \qquad 2\partial_- \widetilde{\psi}_R^1 = m_\psi^1 e^{\eta + i\phi_1}\widetilde{\psi}_L^1, \tag{21}$$

$$-2\partial_+ \widetilde{\psi}_L^1 = m_\psi^1 e^{2\eta - i\phi_1}\widetilde{\psi}_R^1 + 2i\left(\frac{m_\psi^2 m_\psi^3}{im_\psi^1}\right)^{1/2} e^\eta(-\psi_L^2 \widetilde{\psi}_R^3 e^{-i\phi_3} - \widetilde{\psi}_L^3 \psi_R^2 e^{i\phi_2}), \tag{22}$$

$$-2\partial_+ \widetilde{\psi}_L^2 = m_\psi^2 e^{2\eta - i\phi_2}\widetilde{\psi}_R^2 + 2i\left(\frac{m_\psi^1 m_\psi^3}{im_\psi^2}\right)^{1/2} e^\eta(\psi_L^1 \widetilde{\psi}_R^3 e^{-i\phi_3} + \widetilde{\psi}_L^3 \psi_R^1 e^{i\phi_1}), \tag{23}$$

$$2\partial_- \widetilde{\psi}_R^2 = m_\psi^2 e^{\eta + i\phi_2}\widetilde{\psi}_L^2, \qquad -2\partial_+ \widetilde{\psi}_L^3 = m_\psi^3 e^{\eta - i\phi_3}\widetilde{\psi}_R^3, \tag{24}$$

$$2\partial_- \widetilde{\psi}_R^3 = m_\psi^3 e^{2\eta + i\phi_3}\widetilde{\psi}_L^3 + 2i\left(\frac{m_\psi^1 m_\psi^2}{im_\psi^3}\right)^{1/2} e^\eta(\widetilde{\psi}_R^1 \widetilde{\psi}_L^2 e^{i\phi_2} - \widetilde{\psi}_R^2 \widetilde{\psi}_L^1 e^{i\phi_1}), \tag{25}$$

$$\partial^2 \eta = 0, \tag{26}$$

where $\phi_1 \equiv 2\theta_1 - \theta_2$, $\phi_2 \equiv 2\theta_2 - \theta_1$, $\phi_3 \equiv \theta_1 + \theta_2$. Therefore, one has

$$\phi_3 = \phi_1 + \phi_2 \tag{27}$$

The θ fields are considered to be in general complex fields. In order to define the classical generalized sine-Gordon model we will consider these fields to be real.

Apart from the *conformal invariance* the above equations exhibit the $\left(U(1)_L\right)^2 \otimes$ $\left(U(1)_R\right)^2$ *left-right local gauge symmetry*

$$\theta_a \rightarrow \theta_a + \xi_+^a(x_+) + \xi_-^a(x_-), \quad a = 1,2 \tag{28}$$

$$\widetilde{\nu} \rightarrow \widetilde{\nu}; \qquad \eta \rightarrow \eta \tag{29}$$

$$\psi^i \rightarrow e^{i(1+\gamma_5)\Xi_+^i(x_+)+i(1-\gamma_5)\Xi_-^i(x_-)} \psi^i, \tag{30}$$

$$\widetilde{\psi}^i \rightarrow e^{-i(1+\gamma_5)(\Xi_+^i)(x_+)-i(1-\gamma_5)(\Xi_-^i)(x_-)} \widetilde{\psi}^i, \quad i = 1,2,3; \tag{31}$$

$$\Xi_\pm^1 \equiv \pm\xi_\pm^2 \mp 2\xi_\pm^1, \quad \Xi_\pm^2 \equiv \pm\xi_\pm^1 \mp 2\xi_\pm^2, \quad \Xi_\pm^3 \equiv \Xi_\pm^1 + \Xi_\pm^2.$$

One can get global symmetries for $\xi_\pm^a = \mp\xi_\mp^a = $ constants. For a model defined by a Lagrangian these would imply the presence of two vector and two chiral conserved currents. However, it was found only half of such currents [34]. This is a consequence of the lack of a Lagrangian description for the $sl(3)^{(1)}$ CATM model in terms of the model defining fields. So, the vector current

$$J^\mu = \sum_{j=1}^3 m_\psi^j \bar{\psi}^j \gamma^\mu \psi^j \tag{32}$$

and the chiral current

$$J^{5\mu} = \sum_{j=1}^3 m_\psi^j \bar{\psi}^j \gamma^\mu \gamma_5 \psi^j + 2\partial_\mu(m_\psi^1\theta_1 + m_\psi^2\theta_2) \tag{33}$$

are conserved

$$\partial_\mu J^\mu = 0, \quad \partial_\mu J^{5\mu} = 0 \tag{34}$$

The conformal symmetry is gauge fixed by setting

$$\eta = \text{const.} \tag{35}$$

The off-critical model obtained in this way exhibits the vector and topological currents equivalence [25, 29]

$$\sum_{j=1}^3 m_\psi^j \bar{\psi}^j \gamma^\mu \psi^j \equiv \epsilon^{\mu\nu}\partial_\nu(m_\psi^1\theta_1 + m_\psi^2\theta_2), \quad m_\psi^3 = m_\psi^1 + m_\psi^2, \quad m_\psi^i > 0. \tag{36}$$

Moreover, it has been shown that the soliton type solutions are in the orbit of the vacuum $\eta = 0$.

In the next steps we implement the reduction process to get the cGSG model through a gauge fixing of the ATM theory. The local symmetries (28)-(31) can be gauge fixed through

$$i\bar{\psi}^j\psi^j = iA_j = \text{const.}; \quad \bar{\psi}^j\gamma_5\psi^j = 0. \tag{37}$$

From the gauge fixing (37) one can write the following bilinears

$$\widetilde{\psi}_R^j\psi_L^j + \widetilde{\psi}_L^j\psi_R^j = 0, \quad j = 1,2,3; \tag{38}$$

so, the eqs. (37) effectively comprises three gauge fixing conditions.

It can be directly verified that the gauge fixing (37) preserves the currents conservation laws (34), i.e. from the equations of motion (15)-(26) and the gauge fixing (37) together with (35) it is possible to obtain the currents conservation laws (34).

Taking into account the constraints (37) in the scalar sector, eqs. (15), we arrive at the following system of equations (set $\eta = 0$)

$$\partial^2 \theta_1 = M_\psi^1 \sin\phi_1 + M_\psi^3 \sin\phi_3, \tag{39}$$

$$\partial^2 \theta_2 = M_\psi^2 \sin\phi_2 + M_\psi^3 \sin\phi_3, \quad M_\psi^i \equiv 4A_i\, m_\psi^i, \quad i = 1, 2, 3. \tag{40}$$

Define the fields φ_1, φ_2 as

$$\varphi_1 \equiv a\theta_1 + b\theta_2, \quad a = \frac{4\nu_2 - \nu_1}{3\beta_0\nu_1\nu_2}, \quad d = \frac{4\nu_1 - \nu_2}{3\beta_0\nu_1\nu_2} \tag{41}$$

$$\varphi_2 \equiv c\theta_1 + d\theta_2, \quad b = -c = \frac{2(\nu_1 - \nu_2)}{3\beta_0\nu_1\nu_2}, \quad \nu_1, \nu_2 \in \mathbb{R} \tag{42}$$

Then, the system of equations (39)-(40) written in terms of the fields $\varphi_{1,2}$ becomes

$$\partial^2 \varphi_1 = aM_\psi^1 \sin[\beta_0\nu_1(2\varphi_1 - \varphi_2)] + bM_\psi^2 \sin[\beta_0\nu_2(2\varphi_2 - \varphi_1)] + $$
$$(a + b)M_\psi^3 \sin\beta_0[(2\nu_1 - \nu_2)\varphi_1 + (2\nu_2 - \nu_1)\varphi_2)], \tag{43}$$

$$\partial^2 \varphi_2 = cM_\psi^1 \sin[\beta_0\nu_1(2\varphi_1 - \varphi_2)] + dM_\psi^2 \sin[\beta_0\nu_2(2\varphi_2 - \varphi_1)] + $$
$$(c + d)M_\psi^3 \sin\beta_0[(2\nu_1 - \nu_2)\varphi_1 + (2\nu_2 - \nu_1)\varphi_2)] \tag{44}$$

The system of equations above considered for real fields $\varphi_{1,2}$ as well as for real parameters $M_\psi^i, a, b, c, d, \beta_0$ defines the *classical generalized sine-Gordon model* (cGSG). Notice that this classical version of the GSG model derived from the ATM theory is a submodel of the GSG model (11)-(12), defined in section 2., for the particular parameter values $r = \frac{2\nu_1 - \nu_2}{\nu_3}$, $s = \frac{2\nu_2 - \nu_1}{\nu_3}$ and the convenient identifications of the parameters in the coefficients of the sine functions of the both models.

The spinor sector in view of the gauge fixing (37) can be parameterized conveniently as

$$\begin{pmatrix} \psi_R^j \\ \psi_L^j \end{pmatrix} = \begin{pmatrix} \sqrt{A_j/2}\, u_j \\ i\sqrt{A_j/2}\, \frac{1}{v_j} \end{pmatrix}; \quad \begin{pmatrix} \widetilde{\psi}_R^j \\ \widetilde{\psi}_L^j \end{pmatrix} = \begin{pmatrix} \sqrt{A_j/2}\, v_j \\ -i\sqrt{A_j/2}\, \frac{1}{u_j} \end{pmatrix}. \tag{45}$$

Therefore, in order to find the spinor field solutions one can solve the eqs. (17)-(25) for the fields u_j, v_j for each solution given for the cGSG fields $\varphi_{1,2}$ of the system (43)-(44).

3.1. Topological Charges, Baryons as Solitons and Confinement

In this section we will examine the vacuum configuration of the cGSG model and the equivalence between the $U(1)$ spinor and the topological currents (36) in the gauge fixed model and verify that the charge associated to the $U(1)$ current gets confined inside the solitons and kinks of the GSG model; the explicit form of these type of solutions will be obtained in section 5..

It is well known that in 1 + 1 dimensions the topological current is defined as $J^{\mu}_{\text{top}} \sim \epsilon^{\mu\nu}\partial_{\nu}\Phi$, where Φ is some scalar field. Therefore, the topological charge is $Q_{\text{top}} = \int J^{0}_{\text{top}}dx \sim \Phi(+\infty) - \Phi(-\infty)$. In order to introduce a topological current we follow the construction adopted in Abelian affine Toda models, so we define the field

$$\theta = \sum_{a=1}^{2} \frac{2\alpha_a}{\alpha_a^2}\theta_a \qquad (46)$$

where α_a, $a = 1, 2$, are the simple roots of $sl(3,\mathbb{C})$. We then have that $\theta_a = (\theta|\lambda_a)$, where λ_a are the fundamental weights of $sl(3,\mathbb{C})$ defined by the relation [35]

$$2\frac{(\alpha_a|\lambda_b)}{(\alpha_a|\alpha_a)} = \delta_{ab}. \qquad (47)$$

The fields ϕ_j in the equations (15)-(25) written as the combinations $(\theta|\alpha_j)$, $j = 1, 2, 3$, where the $\alpha'_j s$ are the positive roots of $sl(3,\mathbb{C})$, are invariant under the transformation

$$\theta \to \theta + 2\pi\mu \qquad \text{or} \qquad \phi_j \to \phi_j + 2\pi(\mu|\alpha_j), \qquad (48)$$

$$\mu \equiv \sum_{n_a \in \mathbb{Z}} n_a \frac{2\vec{\lambda}_a}{(\alpha_a|\alpha_a)}, \qquad (49)$$

where μ is a weight vector of $sl(3,\mathbb{C})$, these vectors satisfy $(\mu|\alpha_j) \in \mathbb{Z}$ and form an infinite discrete lattice called the weight lattice [35]. However, this weight lattice does not constitute the vacuum configurations of the ATM model, since in the model described by (15)-(26) for any constants $\theta_a^{(0)}$ and $\eta^{(0)}$

$$\psi_j = \widetilde{\psi}_j = 0, \ \theta_a = \theta_a^{(0)}, \ \eta = \eta^{(0)}, \ \widetilde{\nu} = -\mathbf{m}^2 e^{\eta^{(0)}} x^{+}x^{-} \qquad (50)$$

is a vacuum configuration.

We will see that the topological charges of the physical one-soliton solutions of (15)-(26) which are associated to the new fields φ_a, $a = 1, 2$, of the cGSG model (43)-(44) lie on a modified lattice which is related to the weight lattice by re-scaling the weight vectors. In fact, the eqs. of motion (43)-(44) for the field defined by $\varphi \equiv \sum_{a=1}^{2} \frac{2\alpha_a}{\alpha_a^2}\varphi_a$, such that $\varphi_a = (\varphi|\lambda_a)$, are invariant under the transformation

$$\varphi \to \varphi + \frac{2\pi}{\beta_0}\sum_{a=1}^{2}\frac{q_a}{\nu_a}\frac{2\lambda_a}{(\alpha_a|\alpha_a)}, \quad q_a \in \mathbb{Z}. \qquad (51)$$

So, the vacuum configuration is formed by an infinite discrete lattice related to the usual weight lattice by the relevant re-scaling of the fundamental weights $\lambda_a \to \frac{1}{\nu_a}\lambda_a$. The vacuum lattice can be given by the points in the plane $\varphi_1 \times \varphi_2$

$$(\varphi_1, \varphi_2) = \frac{2\pi}{3\beta_0}\left(\frac{2q_1}{\nu_1} + \frac{q_2}{\nu_2}, \frac{q_1}{\nu_1} + \frac{2q_2}{\nu_2}\right), \quad q_a \in \mathbb{Z}. \qquad (52)$$

In fact, this lattice is related to the one in eq. (10) through appropriate parameter identifications. We shall define the topological current and charge, respectively, as

$$J_{\text{top}}^{\mu} = \frac{\beta_0}{2\pi} \epsilon^{\mu\nu} \partial_\nu \varphi, \quad Q_{\text{top}} = \int dx J_{\text{top}}^0 = \frac{\beta_0}{2\pi} [\varphi(+\infty) - \varphi(\infty)]. \tag{53}$$

Taking into account the cGSG fields (43)-(44) and the spinor parameterizations (45) the currents equivalence (36) of the ATM model takes the form

$$\sum_{j=1}^{3} m_\psi^j \bar{\psi}^j \gamma^\mu \psi^j \equiv \epsilon^{\mu\nu} \partial_\nu (\zeta_\psi^1 \varphi_1 + \zeta_\psi^2 \varphi_2), \tag{54}$$

where $\zeta_\psi^1 \equiv \beta_0^2 \nu_1 \nu_2 (m_\psi^1 d + m_\psi^2 b)$, $\zeta_\psi^2 \equiv \beta_0^2 \nu_1 \nu_2 (m_\psi^2 a - m_\psi^1 b)$ and the spinors are understood to be written in terms of the fields u_j and v_j of (45).

Notice that the topological current in (54) is the projection of (53) onto the vector $\frac{2\pi}{\beta_0} (\zeta_\psi^1 \lambda_1 + \zeta_\psi^2 \lambda_2)$.

As mentioned above the gauge fixing (37) preserves the currents conservation laws (34). Moreover, the cGSG model was defined for the off critical ATM model obtained after setting $\eta = \text{const.} = 0$. So, for the gauge fixed model it is expected to hold the currents equivalence relation (36) written for the spinor parameterizations u_j, v_j and the fields $\varphi_{1,2}$ as is presented in eq. (54). Therefore, in order to verify the $U(1)$ current confinement it is not necessary to find the explicit solutions for the spinor fields. In fact, one has that the current components are given by relevant partial derivatives of the linear combinations of the field solutions, $\varphi_{1,2}$, i.e. $J^0 = \sum_{j=1}^{3} m_\psi^j \bar{\psi}^j \gamma^0 \psi^j = \partial_x (\zeta_\psi^1 \varphi_1 + \zeta_\psi^2 \varphi_2)$ and $J^1 = \sum_{j=1}^{3} m_\psi^j \bar{\psi}^j \gamma^1 \psi^j = -\partial_t (\zeta_\psi^1 \varphi_1 + \zeta_\psi^2 \varphi_2)$.

It is clear that the charge density related to this $U(1)$ current can only take significant values on those regions where the $x-$derivative of the fields $\varphi_{1,2}$ are non-vanishing. That is one expects to happen with the bag model like confinement mechanism in quantum chromodynamics (QCD). As we will see below the soliton and kink solutions of the GSG theory are localized in space, in the sense that the scalar fields interpolate between the relevant vacua in a limited region of space with a size determined by the soliton masses. The spinor $U(1)$ current gets the contributions from all the three spinor flavors. Moreover, from the equations of motion (17)-(25) one can obtain nontrivial spinor solutions different from vacuum (50) for each set of scalar field solutions φ_1, φ_2. Therefore, the ATM model can be considered as a multiflavor generalization of the two-dimensional hadron model proposed in [10, 11]. In the last reference a scalar field is coupled to a spinor such that the DSG kink arises as a model for hadron and the quark field is confined inside the bag.

In connection to our developments above let us notice that two-dimensional QCD $_2$ has been used as a laboratory for studying the full four-dimensional theory providing an explicit realization of baryons as solitons. In the picture described above a key role has been played by the equivalence between the Noether and topological currents. Moreover, one notices that the SU(n) ATM theory [5, 6] is a $2D$ analogue of the chiral quark soliton model proposed to describe solitons in QCD $_4$ [36], provided that the pseudo-scalars lie in the Abelian subalgebra and certain kinetic terms are supplied for them.

4.   The Bosonized Effective Action of QCD_2 and the GSG Model

The QCD_2 action is written in terms of gauge fields A_μ and fundamental quark fields ψ as

$$S_F[\psi, A_\mu] = \int d^2x \{-\frac{1}{2e_c^2}Tr(F_{\mu\nu}F^{\mu\nu}) - \bar{\psi}^{ai}[(i\not{\partial} + \not{A})]\psi_{ai} + \mathcal{M}_{ij}\bar{\psi}^{ai}\psi_{aj}\}, \quad (55)$$

where a is the color index ($a = 1, 2, ..., N_C$) and i the flavor index ($i = 1, 2, .., N_f$), e_c, with dimension of a mass, is the quark coupling to the gauge fields, the matrix $\mathcal{M}_{ij} = m_i\delta_{ij}$ (m_i being the quark masses) takes into account the quark mass splitting, and $F_{\mu\nu} \equiv \partial_\mu A_\nu - \partial_\nu A_\mu + i[A_\mu, A_\nu]$ is the gauge field strength.

The bosonized action in the strong-coupling limit ($e_c >>$ all m_i) becomes [20, 23]

$$S_{eff}[g] = N_c S[g] + m^2 N_m \int d^2x Tr_f\Big[\mathcal{D}(g + g^\dagger)\Big], \quad (56)$$

where g is a matrix representing $U(N_f)$, $\mathcal{D} = \frac{\mathcal{M}}{m_0}$, m_0 is an arbitrary mass parameter and the effective mass scale m is given by

$$m = [N_c c m_0(\frac{e_c\sqrt{N_F}}{\sqrt{2\pi}})^{\Delta_c}]^{\frac{1}{1+\Delta_c}}, \quad (57)$$

with

$$\Delta_c = \frac{N_c^2 - 1}{N_c(N_c + N_F)}. \quad (58)$$

In (56) $S[g]$ is the WZNW action and N_m stands for normal ordering with respect to m. In the large N_c limit, which we use below to justify the semi-classical approximation, the scale m becomes

$$m = 0.59 N_F^{\frac{1}{4}} \sqrt{N_c e_c m_0}, \quad (59)$$

so, m takes the value $0.77\sqrt{N_c e_c m_0}$ for three flavors. Notice that we first take the strong-coupling limit $e_c \gg$ all m_i, and then take N_c to be large, thus it is different from the 't Hooft limit [19], where $e_c^2 N_c$ is held fixed.

Following the Skyrme model approach it is useful to first ask for classical soliton solutions of the bosonic action which are heavy in the $N_C \rightarrow$large limit. The action (56) is a massive WZNW action and possesses the property that if g is non-diagonal it can not be a classical solution, as after a diagonalization to

$$g_0 = \text{diag}(e^{-i\beta_0\Phi_1(x)}, e^{-i\beta_0\Phi_2(x)}, ... e^{-i\beta_0\Phi_{N_f}(x)}), \quad \sum_i \Phi_i(x) = \phi(x), \quad \beta_0 \equiv \sqrt{\frac{4\pi}{N_C}} \quad (60)$$

it will have lower energy [37]. Thus, the minimal energy solutions of the massive WZNW model are necessarily in a diagonal form. The majority of particles given by (60) are not going to be stable, but must decay into others.

Previous works consider the diagonal form (60) such that the action (56) reduces to a sum of N_f independent ordinary sine-Gordon models, each one for the corresponding Φ_i field and parameters

$$\widetilde{m}_i^2 = \frac{m_i}{m_0}m^2. \tag{61}$$

In this approach the lowest lying baryon is represented by the minimum-energy configuration for this class of ansatz, i.e.

$$\hat{g}_0(x) = \mathrm{diag}\left(1, 1, ..., e^{-i\sqrt{\frac{4\pi}{N_C}}\Phi_{N_f}}\right), \tag{62}$$

with m_{N_f} chosen to be the smallest mass.

In this paper we will consider the ansatz (60) for

$$N_f = \frac{n}{2}(n-1), \quad N_f \equiv \text{number of positive roots of } su(n), \tag{63}$$

such that $\frac{(n-2)(n-1)}{2}$ linear constraints are imposed on the fields Φ_i. This model corresponds to the generalized sine-Gordon model (GSG) recently studied in the context of the bosonization of the so-called generalized massive Thirring model (GMT) with N_f fermion species [5, 6, 7]. The classical GSG model and some of its properties, such as the algebraic construction based on the affine $sl(n,\mathbb{C})$ Kac-Moody algebra and the soliton spectrum has been the subject of a recent paper [9].

The WZ term in (56) vanishes for either static or diagonal solution, so, for the ansatz (60) and after redefining the additive constant term the action becomes

$$S[g_0] = \int d^2x \sum_{i=1}^{N_f} \left[\frac{1}{2}(\partial_\mu\Phi_i)^2 + 2\widetilde{m}_i^2\left(\cos\beta_0\Phi_i - 1\right)\right], \tag{64}$$

with coupling β_0 and mass parameters $\widetilde{m}_i$ defined in (60) and (61), respectively.

The Φ_i fields in (64) satisfy certain constraints of the type

$$\Phi_p = \sum_{i=1}^{n-1} \sigma_{pi}\Phi_i, \quad p = n, n+1, ..., N_f \tag{65}$$

where σ_{pi} are some constant parameters. From the Lie algebraic construction of the GSG model these parameters arise from the relationship between the positive and simple roots of $su(n)$. Even though our treatment in this section and the section 6. is valid for any N_f, we will concentrate on the $N_f = 3$ case.

It is interesting to recognize the similarity between the potential of the model (64)-(65) for the $N_f = 3$ case [in su(3) GSG model one has $n = N_f = 3$ and just one constraint equation in (65)] and the effective chiral Lagrangian proposed by Witten to describe low-energy behavior of four dimensional QCD [38]. In Witten's approach the potential term reads

$$V^{\text{Witten}}(U) = f_\pi^2\left[-\frac{1}{2}TrM(U + U^\dagger) + \frac{k}{2N_C}(-i\ln\mathrm{Det}\,U - \theta)^2\right], \tag{66}$$

where U is the pseudoscalar field matrix and $M = \mathrm{diag}\left(m_u; m_d; m_s\right)$ is the quark mass matrix. Phenomenologically $m_{\eta'}^2 >> m_\pi^2, m_K^2, m_\eta^2$, implying that $\frac{k}{N_C} > b\,m_s >> b\,m_u, b\,m_d$ [the parameter b is $O(\Lambda)$, where Λ is a hadronic scale]. Because M is diagonal, one can look for a minimum of $V^{\mathrm{Witten}}(U)$ in the form $U = \mathrm{diag}\left(e^{i\phi_1}, e^{i\phi_2}, e^{i\phi_3}\right)$. Since the second term dominates over the first, one has $\sum \phi_j = \theta$ up to the first approximation. So, choosing $\theta = 0$, (66) reduces to a model of type (64)-(65) defined for $N_f = 3$. This is the $sl(3)$ GSG model (11)-(12), which possesses soliton and kink type solutions (see section 5.), and will be the main ingredient of our developments in sections 7. and 8..

The potential term in (64) is invariant under

$$\Phi_i \to \Phi_i + \frac{1}{\beta_0} 2\pi N_i, \ (N_i \in \mathbb{Z}). \tag{67}$$

All finite energy configurations, whether static or time-dependent, can be divided into an infinite number of topological sectors, each characterized by a set

$$\left[n_1, n_2, ..., n_{N_f}\right] = \left[(N_1^+ - N_1^-), (N_2^+ - N_2^-), ..., (N_{N_f}^+ - N_{N_f}^-)\right] \tag{68}$$

$$\Phi_i(\pm\infty) = \frac{1}{\beta_0} 2\pi N_i^\pm \tag{69}$$

corresponding to the asymptotic values of the fields at $x = \pm\infty$. The $n_i's$ satisfy certain relationship arising from the constraints (65) and the invariance (67) (some examples are given in section 5. for the soliton and kink type solutions in the $SU(3)$ case).

Conserved charges, corresponding to the vector current $J_{ij}^\mu = \bar{\psi}_i^a \gamma^\mu \psi_j^a$, can be computed as

$$Q^A[g(x)] = \int dx [J_0(\frac{T^A}{2})], \tag{70}$$

where $(\frac{T^A}{2})$ are the $su(n)$ generators and the $U(1)$ baryon number is obtained using the identity matrix instead of $(\frac{T^A}{2})$. For g_0 given in eq. (60) the baryon number of any given flavor j is given by $\mathcal{Q}_B^j = N_c n_j$, so, the total baryon number becomes

$$\mathcal{Q}_B = N_C(n_1 + n_2 + ... + n_{N_f}), \tag{71}$$

and the "hypercharge" is given by

$$\begin{aligned}
Q_Y &= \frac{1}{2}\mathrm{Tr}\int dx \left(J_0 \lambda_{N_f^2-1}\right) \\
&= \frac{1}{2}N_C\left(n_1 + n_2 + ... + n_{N_f-1} - (N_f - 1)n_{N_f}\right)\sqrt{\frac{2}{N_f^2 - N_f}}. \tag{72}
\end{aligned}$$

The total baryon number is clearly an integer multiple of N_C. In the case of (62) they reduce to $\mathcal{Q}_B = N_C$ and $\mathcal{Q}_Y^0 = -\frac{1}{2}\sqrt{2(N_f - 1)/N_F}N_C$, respectively [for $\sqrt{4\pi/N_C}\Phi_{N_f}(+\infty) = 2\pi, \ \Phi_{N_f}(-\infty) = 0$] [20]. We are choosing the convention in which the quarks have baryon number $Q_B = 1$, so the soliton representing a physical baryon has baryon number N_C.

A global $U_V(N_f)$ transformation $\widetilde{g}_0 = A g_0(x) A^{-1}$ is expected to turn on the other charges. Let us introduce

$$A = \begin{pmatrix} z_1^{(1)} & \cdots & z_1^{(N_f)} \\ z_2^{(1)} & \cdots & z_2^{(N_f)} \\ z_{N_f}^{(1)} & \cdots & z_{N_f}^{(N_f)} \end{pmatrix}, \tag{73}$$

$$\sum_{p=1}^{N_f} z_p^{(i)} z_p^{(j)\star} = \delta_{ij}. \tag{74}$$

Now

$$\widetilde{g}_0 = \sum_{j=1}^{N_f} e^{i\beta_j \Phi_j} \mathbf{Z}^{(j)}, \quad \mathbf{Z}_{pq}^{(j)} = z_p^{(j)} z_q^{(j)\star}, \tag{75}$$

The charges with $\widetilde{g}_0$ are

$$(\widetilde{Q}^0)^A = \frac{1}{2} N_C \mathrm{Tr} \sum_i \left(n_i T^A \mathbf{Z}^{(i)} \right) \tag{76}$$

The baryon number is unchanged. The $U(n)$ possible representations will be discussed below in the semi-classical quantization approach.

5. $sl(3, \mathbb{C})$ GSG Model, Solitons and Kinks as Baryons

Here we summarize some properties of the $sl(3, \mathbb{C})$ GSG model written in the form (11)-(12) relevant to our discussions above, such as the soliton and kink spectrum [9, 22]. The discussions make some connection to the QCD$_2$ developments above, such as (multi-) baryon number of solitons and kinks.

In the following we write the 1-soliton(antisoliton) and 1-kink(antikink) type solutions and compute the relevant (multi-)baryon numbers associated to the $U(1)$ symmetry in the context of QCD$_2$.

5.1. One Soliton/Antisoliton Pair Associated to φ_1

The functions

$$\varphi_1 = \frac{4}{\beta_0} \arctan\{d \, \exp[\gamma_1(x - vt)]\}, \quad \varphi_2 = 0, \tag{77}$$

satisfy the system of equations (11)-(12) for the set of parameters

$$\nu_1 = 1/2, \ \delta_1 = 2, \ \delta_2 = 1, \ \nu_2 = 1, \ \nu_3 = 1, \ r = 1. \tag{78}$$

provided that

$$13\mu_3 = 5\mu_2 - 4\mu_1, \quad \gamma_1^2 = \frac{\beta_0^2}{13}(6\mu_2 + 3\mu_1). \tag{79}$$

This solution is precisely the sine-Gordon 1-soliton associated to the field φ_1 with mass

$$M_1^{sol} = \frac{8\gamma_1}{\beta_0^2}. \tag{80}$$

From (4) and taking into account the parameters (78) one has the relationships between the GSG fields

$$\Phi_1 = -\Phi_2 = \Phi_3 = \varphi_1 \tag{81}$$

Moreover, from (8)-(9) and (78) one gets the relationships

$$n_1 = -n_2 = n_3 \tag{82}$$

Taking into account the QCD$_2$ motivated formula (71) and the eq. (82) above one can compute the baryon number of the GSG soliton (77) taking $n_1 = 1$

$$\mathcal{Q}_B^{(1)} = N_C, \tag{83}$$

where the superindex (1) refers to the associated φ_1 field nontrivial solution.

5.2.   One Soliton/Antisoliton Pair Associated to φ_2

The functions

$$\varphi_2 = \frac{4}{\beta_0}\arctan\{d\exp[\gamma_2(x - vt)]\}, \quad \varphi_1 = 0 \tag{84}$$

solve the system (11)-(12) for the choice of parameters

$$\nu_1 = 1, \ \delta_1 = 1, \ \delta_2 = 2, \ \nu_2 = 1/2, \ \nu_3 = 1, \ s = 1 \tag{85}$$

provided that

$$13\mu_3 = 5\mu_1 - 4\mu_2, \quad \gamma_2^2 = \frac{\beta_0^2}{13}(6\mu_1 + 3\mu_2). \tag{86}$$

This is the sine-Gordon 1-soliton associated to the field φ_2 with mass

$$M_2^{sol} = \frac{8\gamma_2}{\beta_0^2}. \tag{87}$$

As above from (4) and the set of parameters (85) one has the relationships

$$-\Phi_1 = \Phi_2 = \Phi_3 = \varphi_2. \tag{88}$$

From (8)-(9) and (85) one gets the relationship

$$n_1 = -n_2 = -n_3. \tag{89}$$

So, taking into account the QCD$_2$ motivated formula (71) and the above eq. (89) one computes the baryon number of this GSG soliton taking $n_2 = 1$

$$\mathcal{Q}_B^{(2)} = N_C, \tag{90}$$

where the superindex (2) refers to the associated φ_2 field.

5.3.   1-Soliton/1-Antisoliton Pairs Associated to $\hat{\varphi} \equiv \varphi_1 = \varphi_2$

In the case $\varphi_1 = \varphi_2$ one has the 1-soliton solution $\hat{\varphi}$ of the system (11)-(12) associated to the parameters

$$\nu_1 = 1, \; \delta_1 = -1/2, \; \nu_2 = 1, \; \delta_2 = -1/2, \; \nu_3 = -1/2, \; r = s = 1. \tag{91}$$

One has the 1-soliton

$$\begin{aligned} \varphi_1 &= \varphi_2 \equiv \hat{\varphi}, \\ \hat{\varphi} &= \frac{4}{\beta_0} \arctan\{d \, \exp[\gamma_3(x - vt)]\}, \end{aligned} \tag{92}$$

which requires

$$\gamma_3^2 = \beta_0^2\left(\mu_1 + \frac{1}{2}\mu_3\right), \quad \mu_1 = \mu_2. \tag{93}$$

This is a sine-Gordon 1-soliton associated to both fields $\varphi_{1,2}$ in the particular case when they are equal to each other. It possesses a mass

$$M_3^{sol} = \frac{8\gamma_3}{\beta_0^2}. \tag{94}$$

In view of the symmetry (13) which are satisfied by the parameters (91) and (93) one can think of this solution as doubly degenerated.

As above, from (4) and the set of parameters (91) one has the following relationships

$$\Phi_1 = \Phi_2 = -\Phi_3 = \hat{\varphi}. \tag{95}$$

From (8)-(9) and (91) one gets the relationship

$$-2n_3 = n_1 + n_2. \tag{96}$$

So, taking into account the QCD$_2$ motivated formula (71) and the eq. above (96) one computes the baryon number of this GSG solution taking $n_3 = -1$

$$\mathcal{Q}_B^{(\hat{\varphi})} = N_C, \tag{97}$$

where the superindex refers to the associated $\hat{\varphi}$ field.

5.3.1.   Antisolitons and General N-Solitons

The GSG system (11)-(12) reduces to the usual SG equation for each choice of the parameters (78), (85) and (91), respectively. Then, the N−soliton solutions in each case can be constructed as in the ordinary sine-Gordon model.

Using the symmetry (14) one can be able to construct the *1-antisolitons* corresponding to the soliton solutions (77), (84) and (92) simply by changing their signs $\varphi_a \to -\varphi_a$.

5.4. Mass Splitting of Solitons

It is interesting to write some relationships among the various soliton masses.

i) For $\mu_1 \neq \mu_2$ one has respectively the two 1-solitons, (77) and (84), with masses (80) and (87) related by

$$(M_1^{sol})^2 - (M_2^{sol})^2 = \frac{48 N_C}{\pi}(\mu_2 - \mu_1). \tag{98}$$

ii) For $\mu_1 = \mu_2$, there appears the third soliton solution (92)-(93). Then, taking into account (79), (86), (93), (98) and the third soliton mass (94) we have the relationships

$$M_1^{sol} = M_2^{sol}, \quad M_3^{sol} = \sqrt{3/2}\, M_1^{sol}, \tag{99}$$

$$\gamma_1 = \gamma_2 = \sqrt{2/3}\, \gamma_3, \quad \mu_3 = \frac{1}{13}\mu_1. \tag{100}$$

Notice that in this case $M_3^{sol} < M_1^{sol} + M_2^{sol}$, and the third soliton is stable in the sense that energy is required to dissociate it.

5.5. Kink of the Double Sine-Gordon Model as a Multi-baryon

In the system (11)-(12) we perform the following reduction $\phi \equiv \varphi_1 = \varphi_2$ such that

$$\Phi_1 = \Phi_2, \quad \Phi_3 = q\,\Phi_1, \tag{101}$$

with q being a real number.

Moreover, for consistency of the system of equations (11)-(12) we have

$$\mu_1 = \mu_2, \quad \delta_1 = \delta_2 = q/2, \quad \nu_1 = \nu_2, \quad \nu_3 = \frac{q}{2}\nu_1, \quad r = s = 1. \tag{102}$$

Thus the system of Eqs.(11)-(12) reduces to

$$\partial^2 \Phi_{DSG} = -\frac{\mu_1}{\nu_1}\sin(\nu_1 \Phi_{DSG}) - \frac{\mu_3 \delta_1}{\nu_1}\sin(q\nu_1 \Phi_{DSG}), \quad \Phi_{DSG} \equiv \beta_0 \phi. \tag{103}$$

This is the so-called *two-frequency sine-Gordon* model (DSG) and it has been the subject of much interest in the last decades, from the mathematical and physical points of view.

If the parameter q satisfies

$$q = \frac{n}{m} \in \mathbb{Q} \tag{104}$$

with m, n being two relative prime positive integers, then the potential $\frac{\mu_1}{\nu_1^2}(1 - \cos(\nu_1 \Phi_{DSG})) + \frac{\mu_3}{2\nu_1^2}(1 - \cos(q\nu_1 \Phi_{DSG}))$ associated to the model (103) is periodic with period

$$\frac{2\pi}{\nu_1}m = \frac{2\pi}{q\nu_1}n. \tag{105}$$

Then, as mentioned above the theory (103) possesses topological excitations.

From (4) and the set of parameters (102) one has the relationships

$$\Phi_1 = \Phi_2 = \frac{1}{q}\Phi_3 = \nu_1\phi. \tag{106}$$

And from (8)-(9) and (102) one gets the relationship

$$n_3 = \frac{q}{2}(n_1 + n_2). \tag{107}$$

So, taking into account the QCD_2 motivated formula (71) and the eq. (107) above one computes the baryon number of this DSG solution

$$\mathcal{Q}_B^{(DSG)} = N_C(1 + \frac{2}{q})n_3, \quad n_3 \in \mathbb{Z}, \tag{108}$$

where the superindex (DSG) refers to the associated DSG solution.

In the following we will provide some kink solutions for a particular set of parameters. Consider

$$\nu_1 = 1/2, \ \delta_1 = \delta_2 = 1, \ \nu_2 = 1/2, \ \nu_3 = 1/2 \ \text{and} \ q = 2, \ n = 2, \ m = 1 \tag{109}$$

which satisfy (102) and (104). This set of parameters provide the so-called *double sine-Gordon model* (DSG), such that from (106) and (109) the field configurations satisfy

$$\Phi_1 = \Phi_2 = \frac{1}{2}\Phi_3 = \frac{1}{2}\phi. \tag{110}$$

Its potential $-[4\mu_1(\cos\frac{\Phi_{DSG}}{2} - 1) + 2\mu_3(\cos\Phi_{DSG} - 1)]$ has period 4π and has extrema at $\Phi_{DSG} = 2\pi p_1$, and $\Phi_{DSG} = 4\pi p_2 \pm 2\cos^{-1}[1 - |\mu_1/(2\mu_3)|]$ with $p_1, p_2 \in \mathbb{Z}$; the second extrema exists only if $|\mu_1/(2\mu_3)| < 1$. Depending on the values of the parameters β_0, μ_1, μ_3 the quantum field theory version of the DSG model presents a variety of physical effects, such as the decay of the false vacuum, a phase transition, confinement of the kinks and the resonance phenomenon due to unstable bound states of excited kink-antikink states (see [4] and references therein). The semi-classical spectrum of neutral particles in the DSG theory is investigated in ref. [39]. Let us mention that the DSG model has recently been in the center of some controversy regarding the computation of its semiclassical spectrum, see [4, 40].

A particular solution of (103) for the parameters (109) can be written as

$$\Phi_{DSG} := 4\arctan\left[\frac{1}{d}\frac{1 + h\,\exp[2\gamma(x - vt)]}{\exp[\gamma(x - vt)]}\right] \tag{111}$$

provided that

$$\gamma^2 = \beta_0^2\Big(\mu_1 + 2\mu_3\Big), \quad h = -\frac{\mu_1}{4}, \tag{112}$$

5.5.1. A Multi-baryon and the DSG Kink ($h < 0$, $\mu_i > 0$)

For the choice of parameters $h < 0$, $\mu_i > 0$ in (112) the equation (111) provides

$$\phi = \frac{4}{\beta_0}\arctan\left[\frac{-2|h|^{1/2}}{d}\sinh[\gamma_K(x - vt) + a_0]\right], \quad \gamma_K \equiv \pm\beta_0\sqrt{\mu_1 + 2\mu_3}, \quad (113)$$

$$a_0 = \frac{1}{2}\ln|h|.$$

This is the DSG 1-kink solution with mass

$$M_K = \frac{16}{\beta_0^2}\gamma_K\left[1 + \frac{\mu_1}{\sqrt{2\mu_3(\mu_1 + 2\mu_3)}}\ln(\frac{\sqrt{\mu_1 + 2\mu_3} + \sqrt{2\mu_3}}{\sqrt{\mu_1}})\right]. \quad (114)$$

Since one must have $\frac{\mu_3}{\mu_1} > \frac{1}{2}$ (see below for the range of possible values of these parameters) the potential supports one type of minima and thus there exists only one type of topological kink [3]. So, the DSG model possesses only the topological excitation (113) relevant to our QCD_2 discussion.

One can relate the parameters μ_j in (2) to the mass parameters m_i in the effective lagrangian of QCD_2 in (64). So, for the "physical values" $N_f = 3$ and $e_c = 100$MeV for the coupling and taking into account (57), (59) and (61) one has for large N_C

$$\mu_j = 2\frac{m_j}{m_0}m^2 \approx N_C m_j 124(\text{MeV}), \quad (115)$$

thus, the $\mu_j's$ have dimension $(\text{MeV})^2$.

For the values of the mass parameters μ_1, μ_3 in the range $[10^3, 5 \times 10^4](\text{MeV})^2$ (take $m_1 \approx m_2 \approx 52$ MeV; $m_3 = 4$ MeV, notice that these values satisfy the relationship (86)) one can determine the values of the ratio κ between the kink (114) and the third soliton (94) masses

$$\kappa \equiv \frac{M_K}{M_3^{sol}}, \quad 4 < \kappa < 4.2 \quad (116)$$

The baryon number of this DSG kink solution is obtained from (108) taking $q = 2$, $n_3 = 2$

$$\mathcal{Q}_B^{(K)} = 4N_C, \quad (117)$$

where the superindex (K) refers to the associated DSG kink solution.

The above relations (116)-(117) suggest that the decay of the *kink* to four solitons $\{M_j^{sol}\}$ ($j = 1, 2, 3$) is allowed by conservation of energy and charge, however one can see from the kink dynamics that it is a stable object and its fission may require an external trigger. For similar phenomena in soliton dynamics see ref. [41].

Let us emphasize that the baryons with charges $2n_3N_C$ [set $q = 2$ in (108)] for $n_3 = 1, 2, ...$ are assumed to be bound states of 2, 4, ... "basic" baryons, and so, they would correspond to di-baryon states like *deuteron* ($_1^1H^+$) and the "α particle" ($_2^4He^+$). However, we have not found, for the QCD_2 motivated parameter space (μ_1, μ_3) any kink with baryon number $2N_C$. These 2–baryons are expected to be found in the 2–soliton sectors of the GSG model. Notice that in our formalism the four-baryon appears already for $N_f = 3$ as a DSG kink with topological charge $4N_C$, eq. (117). In the formalism of refs. [20, 42] the multibaryons have baryon number kN_C ($k \leq N_f - 1$), so their $(N_f - 1)$–baryon is the one with the greatest baryon number.

5.6. Configuration with Baryon Number $3N_C$

These solutions do not form stable configurations, nevertheless we describe them for completeness. Let us take $\varphi_1 = \varphi_2$, so one has two 1-soliton solutions $\hat{\varphi}_A$ ($A = 1, 2$) of the system (11)-(12) associated to the parameters

$$\nu_1 = 1, \ \delta_1 = 1/2, \ \nu_2 = 1, \ \delta_2 = 1/2, \ \nu_3 = 1/2, \ r = s = 1. \tag{118}$$

As the first 1-soliton one has

$$\varphi_1 = \varphi_2 \equiv \hat{\varphi}_1, \tag{119}$$

$$\hat{\varphi}_1 = \frac{4}{\beta_0}\arctan\{d\ \exp[\gamma_4(x - vt)]\}, \tag{120}$$

which requires

$$d^2 = 1, \ \ 38\gamma_4^2 = \beta_0^2\Big(25\mu_1 + 13\mu_2 + 19\mu_3\Big) \tag{121}$$

This is a sine-Gordon 1-soliton associated to both fields $\varphi_{1,2}$ in the particular case when they are equal to each other. It possesses a mass

$$M_4^{sol} = \frac{8\gamma_4}{\beta_0^2}. \tag{122}$$

In view of the symmetry (13) we are able to write from (121)

$$d^2 = 1, \ \ 38\gamma_5^2 = 25\mu_2 + 13\mu_1 + 19\mu_3, \tag{123}$$

and then one has another 1-soliton from (119)-(120)

$$\varphi_1 = \varphi_2 \equiv \hat{\varphi}_2, \tag{124}$$

$$\hat{\varphi}_2 = \frac{4}{\beta_0}\arctan\{d\ \exp[\gamma_5(x - vt)]\}. \tag{125}$$

It possesses a mass

$$M_5^{sol} = \frac{8\gamma_5}{\beta_0^2}. \tag{126}$$

Similarly, from (4) and the set of parameters (118) one has the following relationships

$$\Phi_1 = \Phi_2 = \Phi_3 = \hat{\varphi}_A, \ \ A = 1, 2. \tag{127}$$

From (8)-(9) and (118) one gets the relationship

$$2n_3 = n_1 + n_2. \tag{128}$$

So, taking into account the QCD$_2$ motivated formula (71) and the eq. (128) one computes the baryon number of this GSG solution taking $n_3 = 1$

$$\mathcal{Q}_B^{(A)} = 3N_C, \tag{129}$$

where the superindex (A) refers to the associated $\hat{\varphi}_A$ field. Therefore, the both solutions $A = 1, 2$, have the same baryon number in the context of QCD$_2$. The individual soliton solutions (120) and (125) have, respectively, a topological charge N_C, since they are sine-Gordon solitons. Then, the configuration (127) with total charge $3N_C$ is composed of three SG solitons. Therefore, by conservation of energy and topological charge arguments one has that the rest mass of the static configurations $A = 1, 2$, with baryon number $3N_C$ will be, respectively

$$M_{4,5}^{config.} \equiv 3\, M_{4,5}^{sol}, \tag{130}$$

where the masses $M_{4,5}^{sol}$ are given by (122), (126).

Moreover, one can verify the following relationships

$$i)\ M_{4,5}^{config.}\ >\ M_1^{sol} + M_2^{sol}; \qquad\qquad \mu_1 \neq \mu_2, \tag{131}$$

$$ii)\ M_4^{config.}\ =\ M_5^{config.} > M_1^{sol} + M_2^{sol} + M_3^{sol}; \qquad \mu_1 = \mu_2, \tag{132}$$

where the soliton masses M_j^{sol} ($j = 1, 2, 3$) are given by (80), (87), (94), respectively. One observes that the configurations $A = 1, 2$, do not form bound states (bound states would be formed if the inequalities (131)-(132) are reversed), and they may decay into the "basic" set $\{M_1^{sol}, M_2^{sol}\}$ or $\{M_1^{sol}, M_2^{sol}, M_3^{sol}\}$ of solitons, such that the excess energy is transferred to the kinetic energy of the solitons.

6. Semi-classical Quantization and the GSG Ansatz

In order to implement the semi-classical quantization let us consider

$$g(x, t) = A(t)g_0(x)A^{-1}(t), \quad A(t) \in U(N_f) \tag{133}$$

and derive the effective action for $A(t)$ by substituting $g(x, t)$ into the original action. So, following similar steps to the ones developed in [20] one can get

$$\widetilde{S}(g(x, t)) - \widetilde{S}(g_0(x)) = \frac{N_C}{8\pi} \int d^2x \mathrm{Tr}\left(\left[A^{-1}\dot{A},\, g_0 \right]\left[A^{-1}\dot{A},\, g_0^\dagger \right] \right) +$$

$$\frac{N_C}{2\pi} \int d^2x \mathrm{Tr}\{(A^{-1}\dot{A})(g_0^\dagger \partial_x g_0)\} + m^2 \mathrm{Tr} \int d^2x [(\mathcal{D}Ag_0 A^\dagger - \mathcal{D}g_0) + c.c.] \tag{134}$$

The action above for $\mathcal{D}_{ij} = \delta_{ij}$ (in this case the last integrand after taking the trace operation vanishes identically) is invariant under global $U(N_f)$ transformation

$$A \to UA, \tag{135}$$

where $U \in G = U(N_f)$. This corresponds to the invariance of the original action (with mass of the same magnitude for all flavors) under $g \to UgU^{-1}$. It is also invariant under the local changes

$$A(t) \to A(t)V(t), \tag{136}$$

where $V(t) \in H$. This subgroup H of G is nothing but the invariance group of g_0. Below we will find some particular cases of H.

We define the Lie algebra valued variables q^i, y_a through $A^{-1}\dot{A} = i\sum\{\dot{q}_i E_{\alpha_i} + \dot{y}_a H^a\}$ in the generalized Gell-Mann representation [43]. In terms of these variables the action (134), for a diagonal mass matrix such that $\mathcal{D}_{ij} = \delta_{ij}$, takes the form

$$S[q, y] = \int dt\{\sum_{i=1}^{N_f} \frac{1}{2M_i}\dot{q}_i \dot{q}_{-i} - \sum_{a=1}^{N_f-1} \sqrt{\frac{2}{(a+1)^2 - (a+1)}} \times$$
$$\times (n_1 + n_2 + ... + n_a - an_{a+1})\dot{y}_a\}, \tag{137}$$

where $q_{\pm i}$ are associated to the positive and negative roots, respectively, and

$$\frac{1}{2M_i} = \frac{N_C}{2\pi}\int_{-\infty}^{+\infty}[1 - \cos\beta_0\Phi_i], \quad \Phi_i \neq 0. \tag{138}$$

In the case of vanishing $\Phi_j \equiv 0$ for a given j one must formally set $M_j = +\infty$ in the relevant terms throughout.

In the case of $\hat{g}_0 = \text{diag}(1, 1, ..., e^{i\beta_{N_f}\Phi_{N_f}})$ the second summation in (137) reduces to the unique term $[-N_c\sqrt{\frac{2(N_f-1)}{N_f}} \dot{y}_{N_f-1}]$ [20].

When written in terms of the general diagonal field $g_0(x)$ and the $U(N_f)$ field $A(t)$, the charges associated to the global $U(N_f)$ symmetry, (70), become

$$Q^B = i\frac{N_C}{8\pi}\int dx \text{Tr}\{T^B A\{(g_0^\dagger\partial_x g_0 - g_0\partial_x g_0^\dagger) + [g_0, [A^{-1}\dot{A}, g_0^\dagger]]\}A^{-1}\} \tag{139}$$

A convenient parameterization, instead of the parameters used in (137), is (73) since in the above expressions, for Q^B and the action (134), there appear the fields A, A^{-1}, as well as their time derivatives. Now, for a diagonal mass matrix such that $\mathcal{D} = \frac{m_i}{m_0}\delta_{ij}$, the expression (134) can be written in terms of the variables $z_p^{(i)}$ subject to the relationships (74)

$$\widetilde{S}(g(x, t)) - \widetilde{S}(g_0(x)) = S[z_p^{(i)}(t), \Phi_i(x)] \tag{140}$$

$$S[z_p^{(i)}(t), \Phi_i(x)] = \frac{N_C}{2\pi}\int d^2x \sum_{p,q;i<j}[\cos(\beta_i\Phi_i - \beta_j\Phi_j) - 1][\dot{z}_p^{(i)}z_q^{(i)\star} \dot{z}_q^{(j)}z_p^{(j)\star}] -$$
$$i\frac{N_C}{2\pi}\int d^2x \sum_{i,p}\beta_i\partial_x\Phi_i\dot{z}_p^{(i)} z_p^{(i)\star} +$$
$$\int dt\, 2[\sum_{i,p}\cos(\beta_i\Phi_i)\widetilde{m}_p^2 z_p^{(i)}z_p^{(i)\star} - \sum_i\cos(\beta_i\Phi_i)\widetilde{m}_i^2] \tag{141}$$

Let us choose the index k corresponding to the smallest mass m_k. So, integrating over x in (141) we may write

$$
\begin{aligned}
S[z_p^{(i)}(t)] &= -\frac{1}{2} \int dt \sum_{i<j}^{N_f} \sum_{p,q}^{N_f} M_{ij}^{-1} \dot{z}_p^{(i)} z_q^{(i)\star} \dot{z}_q^{(j)} z_p^{(j)\star} - \\
&\quad i\frac{N_C}{2} \int dt \sum_i n_i \left[\dot{z}_p^{(i)} z_p^{(i)\star} - z_p^{(i)} \dot{z}_p^{(i)\star} \right] - \\
&\quad \frac{2\pi}{N_c} \int dt \left\{ \sum_{i,p} \left[\frac{\widetilde{m}_p^2}{M_i} - \frac{\widetilde{m}_k^2}{M_k} \right] z_p^{(i)} z_p^{(i)\star} + \frac{2\pi}{N_c} \left[\sum_i \frac{\widetilde{m}_i^2}{M_i} - \frac{\widetilde{m}_k^2}{M_k} N_\phi \right] \right\} \\
&\quad + \int dt (z_p^{(i)} z_p^{(j)\star} - \delta_{ij}) \lambda_{ij}
\end{aligned}
\tag{142}
$$

where N_ϕ is the number of nonvanishing Φ_i fields and we have introduced some Lagrange multipliers enforcing the relationships (74). The constants M_{ij} above are defined by

$$
\frac{1}{2M_{ij}} \equiv \frac{N_C}{2\pi} \int dx [1 - \cos(\beta_0 \Phi_i - \beta_0 \Phi_j)]; \quad i < j.
\tag{143}
$$

If the field solutions are such that $\Phi_i = \Phi_j$, then one must set formally $M_{ij} \to +\infty$ in place of the corresponding constants.

Likewise, we can write the $U(N_f)$ charges, eq. (139), in terms of the $z_p^{(i)}$ variables

$$
\begin{aligned}
Q^A &= \frac{1}{2} T_{\beta\alpha}^A Q_{\alpha\beta}, \\
Q_{\alpha\beta} &= N_C \sum_j n_j z_\alpha^{(j)} z_\beta^{(j)\star} - \frac{i}{2} \sum_{i,j} M_{ij}^{-1} z_\alpha^{(j)} z_\gamma^{(j)\star} \dot{z}_\gamma^{(i)} z_\beta^{(i)\star}.
\end{aligned}
\tag{144}
$$

The second $U(N_f)$ Casimir operator is obtained from the charge matrix elements $Q_{\alpha\beta}$

$$
\begin{aligned}
Q^A Q_A &= \frac{1}{2} Q_{\alpha\beta} Q_{\beta\alpha}, \\
&= \frac{1}{2} N_C^2 \sum_i n_i n_i - \frac{1}{4} \sum_{i<j} \left(M_{ij}^{-1} \right)^2 \dot{z}_\alpha^{(j)} z_\beta^{(j)\star} \dot{z}_\beta^{(i)} z_\alpha^{(i)\star}
\end{aligned}
\tag{145}
$$

The expressions above greatly simplify in certain particular cases of the ansatz (60), the ansatz (62) has been studied extensively in the literature before. In the next subsection we review this case and in further sections we analyze the semiclassical quantization of the GSG ansatz given for $N_f = 3$ flavors.

6.1. Review of Usual Sine-Gordon Soliton and Baryons in QCD$_2$

In this subsection we briefly review the formalism applied to the ansatz (62), which is related to the usual SG one-soliton as the lowest baryon state. In order to calculate the quantum correction it is allowed the sine-Gordon soliton to rotate in $SU(N_f)$ space by a time dependent matrix $A(t)$ as in (133). Let us consider the single baryon state defined for the ansatz (62) for the sine-Gordon soliton solution $\Phi_{N_f} \equiv \Phi_{1-soliton}$ [$\Phi_{1-soliton}$ is given by eq. (1)]; so, in the relations above one must set

$$n_{N_f} = 1; \quad n_j = 0 \, (j \neq N_f); \quad M_{jk}^{-1} \equiv 0 \, (j < k < N_f); \quad M_{jN_f}^{-1} \equiv M_{N_f}^{-1} \, (j < N_f) \quad (146)$$

where $M_{N_f}^{-1}$ can be computed using eq. (138) for $i = N_f$ for the soliton (1)

$$\frac{1}{2M_{N_f}} = \frac{1}{\sqrt{2}\,\widetilde{m}}\left(\frac{N_C}{\pi}\right)^{3/2} \quad (147)$$

Then, for the ansatz (62), i.e. $\hat{g}_0(x) = \text{diag}\left(1, 1, ..., e^{-i\sqrt{\frac{4\pi}{N_C}}\Phi_{N_f}}\right)$, the effective action (142) can be written as

$$
\begin{aligned}
S[z_j^{(N_f)}(t)] = {} & \frac{1}{2M_{N_f}} \int dt [\dot{z}_j^{(N_f)\star} \dot{z}_j^{(N_f)} - (z_i^{(N_f)\star} \dot{z}_i^{(N_f)})(\dot{z}_k^{(N_f)\star} z_k^{(N_f)})] \\
& - \frac{2\pi}{M_{N_f} N_C} \int dt \sum_{i=1}^{N_f} \left(\widetilde{m}_i^2 - \widetilde{m}_{N_f}^2\right) z_i^{(N_f)\star} z_i^{(N_f)} \\
& - i\frac{N_C}{2} \int dt\, n_{N_f} \left(z_j^{(N_f)\star} \dot{z}_j^{(N_f)} - \dot{z}_j^{(N_f)\star} z_j^{(N_f)}\right) \\
& + \int dt [(z_p^{(N_f)} z_q^{(N_f)\star} - \delta_{pq})\lambda^{pq}], \quad (148)
\end{aligned}
$$

where $n_{N_f} = 1$, M_{N_f} is given by (147) and m_{N_f} entering $\widetilde{m}_{N_f}$ is chosen to be the smallest quark mass. Notice that for equal quark masses the second line in eq. (148) vanishes identically. According to (135)-(136), the symmetries of $S[z_j^{(N_f)}(t)]$ are the global $U(N_f)$ group (for equal quark masses) under which

$$z_\alpha^{(N_f)} \to z_\alpha^{'(N_f)} = U_{\alpha\beta} z_\beta^{(N_f)}, \quad U \in U(N_f), \quad (149)$$

and a local $U(1)$ subgroup of H under which

$$z_\alpha^{(N_f)} \to z_\alpha^{'(N_f)} = e^{i\delta(t)} z_\alpha^{(N_f)}. \quad (150)$$

The action (148) has been considered in order to find the quantum correction to the soliton mass for certain representations R of the flavor symmetry $SU(N_f)$. The case of equal quark masses has been studied in the literature [20, 21, 24, 44]. Certain properties in the case of different quark masses have been considered in [23] for the ansatz (62).

In this approach the minimum-energy configuration for the class of ansatz (62), with m_{N_f} the smallest mass, corresponds to the state of lowest-lying baryon [20] which in the large-N_C limit possesses the classical mass

$$M_{baryon}^{cl} = 4\widetilde{m}_{N_f}\left(\frac{2N_C}{\pi}\right)^{1/2} \approx 1.90 N_f^{1/4} \sqrt{e_c m_{N_f}} N_C, \quad (151)$$

where $\widetilde{m}_{N_f}$ has been given in (61) for $i = N_f$.

Moreover, for the Ansatz (62) the $SU(N_f)$ charges become

$$Q_{\alpha\beta} = N_C n_{N_f} z_\alpha^{(N_f)} z_\beta^{(N_f)\star} + \frac{i}{2M_{N_f}} \left[z_\alpha^{(N_f)} z_\beta^{(N_f)\star} (\dot{z}_\delta^{(N_f)} z_\delta^{(N_f)\star} - z_\delta^{(N_f)} \dot{z}_\delta^{(N_f)\star}) + \right.$$
$$\left. z_\alpha^{(N_f)} \dot{z}_\beta^{(N_f)\star} - \dot{z}_\alpha^{(N_f)} z_\beta^{(N_f)\star} \right] \tag{152}$$

The corresponding second Casimir can be obtained from (145)

$$Q_A Q^A = \frac{1}{2} Q_{\alpha\beta} Q_{\beta\alpha} = \frac{1}{2} N_C^2 n_{N_f}^2 + \frac{1}{4M_{N_f}^2} \left(Dz \right)_\alpha^\dagger \left(Dz \right)_\alpha, \quad Dz \equiv \dot{z} - z(z^\dagger \dot{z}) \tag{153}$$

Moreover, denoting the $SU(N_f)$ second Casimir operator by $C_2(N_f)$ one can write

$$Q_A Q^A = C_2(N_f) + \frac{1}{2N_f} (\mathcal{Q}_B)^2, \tag{154}$$

where $\mathcal{Q}_B$ is the baryon number (71), which in this case reduces to $\mathcal{Q}_B = N_C$.

In the case of a single baryon given by $\hat{g}_0$, eq. (62), and for unequal quark masses, the hamiltonian is linear in the quadratic Casimir operator. To see this we now derive the hamiltonian corresponding to the action (148). The canonical momenta are given by

$$p_\alpha = \frac{\partial L}{\partial \dot{z}_\alpha^{(N_f)\star}} = \frac{1}{2M_{N_f}} \left[\dot{z}_\alpha^{(N_f)} - \left(\dot{z}_\beta^{(N_f)} z_\beta^{(N_f)\star} \right) z_\alpha^{(N_f)} \right] + \frac{iN_C}{2} z_\alpha^{N_f} \tag{155}$$

and there is a conjugate expression for p_α. Therefore, from $H = p_\alpha \dot{z}_\beta^{(N_f)\star} + p_\alpha^\star \dot{z}_\beta^{(N_f)} - L$, one can get the hamiltonian

$$H = \frac{1}{2M_{N_f}} \left(Dz \right)_\alpha^\dagger \left(Dz \right)_\alpha + \frac{2\pi}{M_{N_f} N_C} \sum_{i=1}^{N_f} \left(\widetilde{m}_i^2 - \widetilde{m}_{N_f}^2 \right) z_i^{(N_f)\star} z_i^{(N_f)}. \tag{156}$$

However, one must take a proper care of the relevant constraint (74) which was incorporated through the addition of a Lagrange multiplier in the action (148). A proper treatment of a constrained system must be performed at this point [20]. In [20, 44] it was shown that the local $U(1)$ gauge symmetry (150) leads to the constraint

$$Q_{N_f N_f} = 0 \Rightarrow \mathcal{Q}_B = \sqrt{2N_f(N_f - 1)} Q_Y \tag{157}$$

which has to be imposed on physical states. This implies that the representation R must contain a state with Y charge

$$\bar{Q}_Y = \sqrt{\frac{1}{2N_f(N_f - 1)}} N_C. \tag{158}$$

The remaining states will be generated through the application of the $SU(N_f)$ transformations to this one. For states with only quarks and no antiquarks, the condition that $\mathcal{Q}_B = N_C$ implies that only representations described by Young tableaux with N_C boxes appear. The additional constraint that $Q_Y = \bar{Q}_Y$ implies that all N_C quarks belong to

$SU(N_f - 1)$, i.e., this state does not involve the $N_f'^{th}$ quark flavor. These constraints are automatically satisfied in the totally symmetric representation of N_C boxes, which is the only representation possible in two dimensions. This is because the state wave functions have to be constructed out of the components of the complex vector $z^{(N_f)}$ as

$$\psi(z^{(N_f)}, z^{(N_f)\star}) = (z_1^{(N_f)})^{p_1}...(z_{N_f}^{(N_f)})^{p_{N_f}}(z_1^{(N_f)\star})^{q_1}...(z_{N_f}^{(N_f)\star})^{q_{N_f}} \tag{159}$$

with $\sum_{i=1}^{N_f}(p_i - q_i) = N_C$.

The lowest such multiplet has

$$\sum_{i=1}^{N_f} p_i = N_C \quad \text{and all} \quad q_i = 0 \tag{160}$$

This multiplet corresponds to the Young tableaux

$$\overbrace{\square\cdots\square}^{N_c} \tag{161}$$

In QCD$_2$ for $N_C = 3$, $N_F = 3$ we get only the **10** of $SU(3)$.

Then, taking into account (153), (154) and (61), the expression (156) becomes [20, 23]

$$H = M^{cl}_{baryon}\left\{1 + \left(\frac{\pi}{2N_C}\right)^2\left[C_2(R) - \frac{n_{N_f}^2 N_C^2}{2N_f}(N_f - 1)\right] + \sum_{i=1}^{N_f}\frac{m_i - m_{N_f}}{m_{N_f}}|z_i^{(N_f)}|^2\right\}, \tag{162}$$

where M^{cl}_{baryon} is given by (151) and $C_2(R)$ is the value of the quadratic Casimir for the flavor representation R of the baryon. For a baryon state given by SG 1-soliton solution one must set $n_{N_f} = 1$ in the hamiltonian above. Notice that the Hamiltonian depends on m_0 only through M^{cl}_{baryon}, so the overall mass scale is undetermined, only the mass ratios are meaningful. The mass term contributions come from quantum fluctuations around the classical soliton, consistency with the semi-classical approximation requires that it be very small compared to one. However, these terms vanish for equal quark masses [20, 21]. The **10** baryon mass becomes

$$M(baryon) = M_{classical}\left[1 + \left(\frac{\pi^2}{8}\right)\frac{N_f - 1}{N_C}\right]. \tag{163}$$

Notice that the quantum correction is suppressed by a factor of N_C. Moreover, the quantum correction for $N_C = 3$, $N_f = 3$ numerically becomes ~ 0.82.

The hamiltonian (162) taken for equal quark masses has been used to compute the energy of the first exotic baryon $\mathcal{E}_1$(a state containing $N_C + 1$ quarks and just one anti-quark) by taking the corresponding Casimir $C_2(\mathcal{E}_1)$ for $R = \mathbf{35}$ of flavor relevant to the exotic state [21]. For further analysis we record the mass of this exotic baryon

$$M(\mathcal{E}_1) = M(classical)\left[1 + \frac{\pi^2}{8}\frac{1}{N_C}\left(3 + N_f - \frac{6}{N_f}\right) + \frac{3\pi^2}{8}\frac{1}{N_C^2}\left(N_f - \frac{3}{N_f}\right)\right]. \tag{164}$$

In the interesting case $N_C = 3$, $N_f = 3$ this becomes

$$M(\mathbf{35}) = M(classical)\left\{1 + \frac{\pi^2}{4}\right\}. \tag{165}$$

In this case the correction due to quantum fluctuations around the classical solution is larger than the classical term. So, the semi-classical approximation may not be a good approximation. However, observe that the ratio $M(\mathbf{35})/M(\mathbf{10}) \sim 1.9$, which is 17% larger than the ratio between the experimental masses of the Θ^+ and the nucleon. See more on this point below. These semi-classical approximations may be improved by introducing different ansatz for g_0 and considering unequal quark mass parameters. These points will be tackled in the next sections.

7. The GSG Model, the Unequal Quark Masses and Baryon States

In the following we will concentrate on the effective action (142) for the particular case $N_f = 3$ and unequal quark mass parameters. So, the $SU(3)$ flavor symmetry is broken explicitly by the mass terms.

The effective Lagrangian in the case of $N_f = 3$ from (142), upon using (74), can be written as

$$
\begin{aligned}
S[z_p^{(i)}(t)] \;=\; & \frac{1}{4}\int dt \Big\{ \left(M_{12}^{-1} + M_{13}^{-1} - M_{23}^{-1}\right)\Big[\dot{z}_\alpha^{(1)}\dot{z}_\alpha^{(1)\star} - \dot{z}_\alpha^{(1)}z_\alpha^{(1)\star}z_\beta^{(1)}\dot{z}_\beta^{(1)\star}\Big] + \\
& \left(M_{12}^{-1} - M_{13}^{-1} + M_{23}^{-1}\right)\Big[\dot{z}_\alpha^{(2)}\dot{z}_\alpha^{(2)\star} - \dot{z}_\alpha^{(2)}z_\alpha^{(2)\star}z_\beta^{(2)}\dot{z}_\beta^{(2)\star}\Big] + \\
& \left(-M_{12}^{-1} + M_{13}^{-1} + M_{23}^{-1}\right)\Big[\dot{z}_\alpha^{(3)}\dot{z}_\alpha^{(3)\star} - \dot{z}_\alpha^{(3)}z_\alpha^{(3)\star}z_\beta^{(3)}\dot{z}_\beta^{(3)\star}\Big]\Big\} - \\
& i\frac{N_C}{2}\int dt \sum_{i,p} n_i\Big[\dot{z}_p^{(i)}z_p^{(i)\star} - z_p^{(i)}\dot{z}_p^{(i)\star}\Big] - \\
& \int dt\Big\{\frac{2\pi}{N_c}\sum_{i,p}\Big[\frac{\tilde{m}_p^2}{M_i} - \frac{\tilde{m}_k^2}{M_k}\Big]z_p^{(i)}z_p^{(i)\star} + \frac{2\pi}{N_c}\Big[\sum_i \frac{\tilde{m}_i^2}{M_i} - \frac{\tilde{m}_k^2}{M_k}N_\phi\Big]\Big\}
\end{aligned} \tag{166}
$$

From (145) and following similar steps the second $U(3)$ Casimir operator can be written as

$$
\begin{aligned}
Q^A Q_A \;=\; & \frac{1}{2}Q_{\alpha\beta}Q_{\beta\alpha}, \\
=\; & \frac{1}{2}N_C^2 \sum_j n_j n_j + \\
& \frac{1}{8}\Big\{\left(M_{12}^{-2} + M_{13}^{-2} - M_{23}^{-2}\right)\Big[\dot{z}_\alpha^{(1)}\dot{z}_\alpha^{(1)\star} - \dot{z}_\alpha^{(1)}z_\alpha^{(1)\star}z_\beta^{(1)}\dot{z}_\beta^{(1)\star}\Big] + \\
& \left(M_{12}^{-2} - M_{13}^{-2} + M_{23}^{-2}\right)\Big[\dot{z}_\alpha^{(2)}\dot{z}_\alpha^{(2)\star} - \dot{z}_\alpha^{(2)}z_\alpha^{(2)\star}z_\beta^{(2)}\dot{z}_\beta^{(2)\star}\Big] + \\
& \left(-M_{12}^{-2} + M_{13}^{-2} + M_{23}^{-2}\right)\Big[\dot{z}_\alpha^{(3)}\dot{z}_\alpha^{(3)\star} - \dot{z}_\alpha^{(3)}z_\alpha^{(3)\star}z_\beta^{(3)}\dot{z}_\beta^{(3)\star}\Big]\Big\}.
\end{aligned} \tag{167}
$$

As a particular case for the ansatz (62) let us take $N_f = 3$, so $n_1 = n_2 = 0$ in (68). In (143) one can set formally $M_{12} \equiv +\infty$ and in view of (138) the remaining parameters can be written as $M_{13} = M_{23} \equiv M_3$. Thus, taking into account these parameters the expressions for the action (166) and the second Casimir (167) reduce to the well known ones (148) and (153), respectively.

Next, we discuss the action (166) and the second Casimir (167) operator for the soliton and kink type solutions of the GSG model. In subsections 5.1., 5.2. and 5.3. we classify these type of solutions. There are three $1-$soliton solutions [see eqs. (77), (84) and (92)] which correspond to baryon number N_C [see eqs. (83), (90) and (97)], because the GSG model possesses the symmetry (13) the third soliton is doubly degenerated. From the fields relationships (81), (88) and (95) one has the three $1-$soliton cases

$$i)\Phi_1 = -\Phi_2 = \Phi_3 = \varphi_1 \quad \Rightarrow \quad M_{13} = +\infty, \; M_{12} = M_{23} \equiv \mathcal{M}_2;$$
$$M_1 = M_2 = M_3 = \widetilde{M_2}, \tag{168}$$

$$ii) -\Phi_1 = \Phi_2 = \Phi_3 = \varphi_2 \quad \Rightarrow \quad M_{23} = +\infty, \; M_{12} = M_{13} \equiv \mathcal{M}_1;$$
$$M_1 = M_2 = M_3 = \widetilde{M_1}, \tag{169}$$

$$iii)\Phi_1 = \Phi_2 = -\Phi_3 = \hat{\varphi} \quad \Rightarrow \quad M_{12} = +\infty, \; M_{13} = M_{23} \equiv \mathcal{M}_3;$$
$$M_1 = M_2 = M_3 = \widetilde{M_3} \tag{170}$$

where the eqs. (143) and (138) have been used, respectively, to define the parameters $\mathcal{M}_j$ and $\widetilde{M_j}$ in the right hand sides of the relationships above.

In section 5. we record the kink type solution [see eq. (113)] which corresponds to the GSG reduced model called double sine-Gordon theory. This solution corresponds to baryon number $4N_C$ [see eq. (117)]. Thus, from (110), (143) and (138) one has

$$\Phi_1 = \Phi_2 = \frac{1}{2}\Phi_3 = \frac{1}{2}\phi \quad \Rightarrow \quad M_{12} = +\infty, \; M_{13} = M_{23} \equiv \mathcal{M}_K;$$
$$M_1 = M_2 \equiv \mathcal{M}_K, \; M_3 \equiv \mathcal{M}_{2K} \tag{171}$$

The solutions with baryon numbers $2N_C$ and $3N_C$ correspond to composite configurations formed by multi-solitons of the GSG model. These states (i.e. multi-baryons) deserve a careful treatment which we hope to undertake in future.

7.1.   GSG Solitons and the States with Baryon Number N_C

For the particular cases (168)-(170) one can rewrite the action (166) such that for each case the terms quadratic in time derivatives reduce to a term depending only on one variable, say $z_i^{(\hat{l})}$, related to the $\hat{l}$'th column of the matrix A. The reason is that the symmetries of the quantum mechanical lagrangian and actual manifold on which $A(t)$ lives depend on the properties of the ansatz g_0. For the ansatz g_0 related to the GSG model one can see that the space-time dependent field g in eq. (133) can be rewritten only in terms of certain columns of A. For example, in the case (170) above the matrix $g(x, t)$ can be written as

$$g_{\alpha\beta}(x, t) = [Ag_0 A^{-1}]_{\alpha\beta}$$
$$= \delta_{\alpha\beta} e^{i\beta_0\hat{\varphi}} - 2i\sin(\beta_0\hat{\varphi})z_\alpha^{(3)}z_\beta^{(3)\star}, \tag{172}$$

which clearly depends only on the third column of A. So, we may think that the left hand side of (134), i.e. $[\widetilde{S}(g(x,t)) - \widetilde{S}(g_0(x))]$, entering the expression of the semi-classical quantization approach, would in principle be written only in terms of the third column of A. However, in order to envisage certain local symmetries it is useful to write the terms first order in time derivatives as depending on the full parameters $z_i^{(j)}$ of the field A. These terms arise from the WZW term and provides the Gauss law type N_z number conservation law [See eq. (179) below]. An additional $SU(2) \in H$ (see (173)) local symmetry will be described below. Moreover, this picture is in accordance with the counting of the degrees of freedom. In fact, the effective action (134) possesses the local gauge symmetry (136), where in the case of field configuration (170) the gauge group H becomes

$$H = SU(2) \times U(1)_B \times U(1)_Y, \tag{173}$$

with the last two $U(1)$ factors related to baryon number and hypercharge, respectively. Thus, the effective action (166) will be an action for the coordination describing the coset space $G/H = SU(3) \times U(1)_B/SU(2) \times U(1)_B \times U(1)_Y = CP^2$. The Φ_i fields and symmetries of g_0 also determine the values and relationships between the parameters M_{ij} in (168)-(170), such that certain coefficients in (166) depending on these parameters vanish identically, thus leaving a subset of $z_i^{(j)}$ variables which must be consistent with the counting of the degrees of freedom. For example this picture is illustrated in the case (170) where the coefficients $(M_{12}^{-1} + M_{13}^{-1} - M_{23}^{-1})$ and $(M_{12}^{-1} - M_{13}^{-1} + M_{23}^{-1})$ vanish identically, leaving an action with kinetic term depending only on the variables $z_\alpha^{(3)}$. However, the mass and WZW terms are conveniently written in terms of the complete $z_i^{(j)}$ variables.

So, for each case in (168)-(170) labeled by $\hat{l}$, the action can be written as

$$S[z_p^{(i)}(t)] = \frac{1}{2} \int dt\, \mathcal{M}_{\hat{l}}^{-1} \left[\dot{z}_\alpha^{(\hat{l})} \dot{z}_\alpha^{(\hat{l})\star} - \dot{z}_\alpha^{(\hat{l})} z_\alpha^{(\hat{l})\star} z_\beta^{(\hat{l})} \dot{z}_\beta^{(\hat{l})\star} \right] - $$
$$i \frac{N_C}{2} \int dt \sum_{i,p} n_i \left[\dot{z}_p^{(i)} z_p^{(i)\star} - z_p^{(i)} \dot{z}_p^{(i)\star} \right] - \frac{2\pi}{N_c \widetilde{M}_{\hat{l}}} \int dt \sum_{i,j} \widetilde{m}_i^2 |z_i^{(j)}|^2. \tag{174}$$

In the relation above we must assign the relevant set of values to the indices n_i ($i = 1, 2, 3$) for the relevant case in (168)-(170). The first term in (174) is the usual CP^2 quantum mechanical action, while the terms first order in time-derivatives are modifications due to the WZ term, as arisen from (134) and (142). Notice that the last term was originated from the unequal quark mass terms.

Following similar steps as in the single baryon case (see eqs. (155)-(156)) one can obtain the hamiltonian

$$H = \frac{1}{2\mathcal{M}_{\hat{l}}} \left(Dz^{(\hat{l})} \right)_\alpha^\dagger \left(Dz^{(\hat{l})} \right)_\alpha + \frac{2\pi}{N_c \widetilde{M}_{\hat{l}}} \sum_{i,j} \widetilde{m}_i^2 |z_i^{(j)}|^2, \tag{175}$$

where $\left(Dz^{(\hat{l})} \right)_\alpha = \dot{z}_\alpha^{(\hat{l})} - z_\alpha^{(\hat{l})}(z_\beta^{(\hat{l})\star} \dot{z}_\beta^{(\hat{l})})$.

Similarly, the corresponding second Casimir becomes

$$Q^A Q_A = \frac{1}{2} Q_{\alpha\beta} Q_{\beta\alpha},$$

$$= \frac{1}{2}N_C^2 \sum_i |n_i|^2 + \frac{1}{4\mathcal{M}_{\hat{l}}^2} \left(Dz^{(l)}\right)_\alpha^\dagger \left(Dz^{(l)}\right)_\alpha \quad (176)$$

Then from (175)-(176) and taking into account $Q^A Q_A = C_2 + \frac{1}{2N_f}\sum_i(\mathcal{Q}_B^i)^2$ one can get

$$H = 2\mathcal{M}_{\hat{l}}\left(C_2 + \frac{1}{2N_f}\sum_i(\mathcal{Q}_B^i)^2 - \frac{1}{2}N_C^2 \sum_i |n_i|^2\right) + \frac{2\pi}{N_c\widetilde{\mathcal{M}}_{\hat{l}}}\sum_{i,j=1}^{3}\widetilde{m}_i^2|z_i^{(j)}|^2 (177)$$

where $\mathcal{Q}_B^i = n_i N_C$ for a convenient choice of the indices n_i, which in the cases (168)-(170) is simply $|n_i| = 1$ [see also eqs. (83), (90) and (97) for 1-soliton configurations]. The parameters $\mathcal{M}_{\hat{l}}$, $\widetilde{\mathcal{M}}_{\hat{l}}$ can be computed for the relevant solitons. They become

$$\frac{1}{2\mathcal{M}_{\hat{l}}} = \frac{1}{\widetilde{m}}\frac{2\sqrt{2}}{3}\left(\frac{N_C}{\pi}\right)^{3/2}, \qquad \frac{1}{2\widetilde{\mathcal{M}}_{\hat{l}}} = \frac{1}{\sqrt{2}\,\widetilde{m}}\left(\frac{N_C}{\pi}\right)^{3/2} \quad (178)$$

Some comments concerning the two hamiltonians (162) and (177) are in order here. Even though they correspond to one baryon state (baryon number N_C) they look different. In fact, the hamiltonian (177) incorporates additional terms. First, due to the ansatz (60) related to the GSG model one has some set of field solutions comprising in total three possibilities (168)-(170) with baryon number N_C, each case being characterized by the set of parameters $\mathcal{M}_{\hat{l}}$, $\widetilde{\mathcal{M}}_{\hat{l}}$ and relevant combinations of the indices n_j which are related to the baryon number of the configuration $\{\Phi_j\}$, $j = 1, 2, 3$. So, the terms $-\frac{N_C^2}{2}$ and $\frac{N_C^2}{2N_f}$ in (162) translate to $-\frac{N_C^2}{2}\sum_i n_i^2$ and $\frac{1}{2N_f}\sum_i(\mathcal{Q}_B^i)^2$, respectively, in the new hamiltonian (177). Second, the mass term expression allows an exact summation due to unitarity, thus giving a constant additional term to the hamiltonian (see below). The corresponding term in (162), obtained in [23], does not permit an exact summation.

7.2. Lowest Lying Baryon State and the GSG Soliton

So far, the treatment for each case (168)-(170) followed similar steps; however, in order to compute the quantum correction to the soliton mass we choose the one from the classification (168)-(170) with the minimum classical energy solution. Thus, taking into account the "physically" motivated inequalities $m_3 < m_1 < m_2$ (or $\mu_3 < \mu_1 < \mu_2$) [eq. (115) relates the μ_j's and the m_j's] one observes that the soliton with mass M_2^{sol} [see eq. (87)] possesses the smallest mass according to the relationship (98). This corresponds to the *second case* (169) classified above; so one must set the index $\hat{l} = 1$ in the action (174).

The variation of the action (174) under $z_\alpha^{(j)} \to e^{i\delta(t)}z_\alpha^{(j)}$ is due to the WZW term: $\Delta S = N_c(n_1 + n_2 + n_3)\int dt\, \dot{\delta}$. This implies

$$N_z = \frac{\Delta S}{\Delta \dot{\delta}} = N_c\left(n_1 + n_2 + n_3\right), \quad (179)$$

which is an analog of the Gauss law, and restricts the allowed physical states [45]. For the soliton configuration with baryon number N_C, (169), under consideration in this subsection, we have $n_1 = -n_2 = -n_3 = -1 \to n_1 + n_2 + n_3 = 1$ [see eq. (89)] implying

$$N_z = N_C. \quad (180)$$

Therefore, for any wave function, written as a polynomial in z and $z^\star$ the number of the z minus the number of the $z^\star$ must be equal to N_C. But due to a larger local symmetry we will have more restrictions. Thus, as commented earlier the (massless part) effective action (174) is invariant under the local $SU(2)$ symmetry. This can be easily seen by defining "local gauge potentials"

$$\widetilde{A}_{\beta\alpha}(t) = -\sum_p z_p^{(\beta)\star} \dot{z}_p^{(\alpha)}, \quad \alpha, \beta = 2, 3. \tag{181}$$

Under the local gauge transformation corresponding to $\Lambda(t)$, one has

$$\widetilde{A}(t) \rightarrow e^{i\Lambda} \widetilde{A} e^{-i\Lambda} + \partial_t e^{i\Lambda} e^{-i\Lambda}. \tag{182}$$

Then we have that the WZW term in (174) for the variables z_p^α, $\alpha = 2, 3$ (take $\hat{l} = 1$, $n_2 = n_3 = 1$) remain invariant under the transformation (182)

$$iN_C \int dt \, \mathrm{Tr} \, \dot{z}_p^{(\alpha)\star} z_p^{(\beta)} \equiv iN_C \int dt \, \mathrm{Tr} \, \widetilde{A} \Rightarrow iN_C \int dt \, \mathrm{Tr} \widetilde{A} \tag{183}$$

Remember that the variables z_p^α do not appear in the kinetic term of (174). The local symmetry above imply that the allowed physical states must be singlets under the $SU(2)$ symmetry in flavor space. So, the wave functions for $z's$ only (analogous to quarks only for QCD) must be of the form

$$\psi_2(z) = \Pi_{i=1}^{N_C} \left(\epsilon_{\alpha_1 \alpha_2} z_{i_1}^{(\alpha_1)} z_{i_2}^{(\alpha_2)} \right), \quad \alpha_1, \alpha_2 = 2, 3, \tag{184}$$

where $1 \leq i_1, i_2 \leq N_f$.

Then, taking into account the restrictions of the types (180) and (184) the most general state can be written as

$$\widetilde{\psi}(z, z^\star) = \psi_2(z) \left[\Pi_{\{p,q\}} (z_p^{(\alpha)\star} z_q^{(\alpha)})^{n_{pq}} \right], \tag{185}$$

and the products are defined for some sets of indices. This wave function generalizes the one given in (159).

Next, let us compute the mass of the state represented for wave functions of the form $\widetilde{\psi}(t) = \psi_2(z) \, \Pi_i (z_i^{(1)})^{p_i}$ where $(\sum_{i=1}^{N_f} p_i = N_C)$.

Combining the hamiltonian (177), the relevant parameters (178) and the classical soliton mass term, for the R baryon we have

$$M(baryon) = M_{classical} \left\{ 1 + \frac{3}{4} (\frac{\pi}{2N_C})^2 \left[C_2(R) - \frac{N_C^2}{2}(N_f - 1) + \frac{1}{2\widetilde{m}^2} \sum_i \widetilde{m}_i^2 \right] \right\}. \tag{186}$$

where

$$M_{classical} = 4\widetilde{m} (\frac{2N_C}{\pi})^{1/2}, \quad \widetilde{m}^2 = \frac{1}{13} (\frac{m^2}{m_0}) \left(6m_1 + 3m_2 \right). \tag{187}$$

The last term in (177) simplifies to a constant term by unitarity condition of the matrix elements $z_i^{(j)}$ and the parameter $\widetilde{m}$ corresponds to the one-soliton parameter once the identification $\gamma_2^2 = 2\beta_0^2 \widetilde{m}^2$ is made in (86) by comparing the SG one-solitons (1) and (84). Even though the computations are explicitly made for $N_f = 3$ it is instructive to leave the number of flavors as a variable. In the case of the **10** baryon one has

$$M(baryon) = M_{classical}\left[1 + \frac{3\pi^2}{32}\frac{N_f - 1}{N_C} - \frac{3\pi^2}{32}\frac{(N_f - 1)^2}{N_f} + \frac{3}{2}\right]. \quad (188)$$

In the following we discuss the correction terms to the earlier expression (163) for the **10** baryon as compared to the last improved expression (188). The quantum correction of (163) is multiplied by $3/4$ and the last two terms in (188) are new contributions due to the GSG ansatz used and the unequal quark mass terms. The last term contribution in (186) was simplified providing a numerical term $3/2$ in (188) thanks to unitarity and the relationship between the quark masses (86) which is a condition to get the relevant soliton solution. This term apparently may not be consistent with a quantum correction around the classical solution since consistency with the semi-classical approximation requires it be small compared to one. However, this term must be combined with the third term which gives a negative value contribution and is an additional term independent of N_C, as is the last numerical $3/2$ term under discussion. In fact, for $N_C = 3$, $N_f = 3$, numerically these two terms contribute ~ 0.27, which is acceptable. The N_C dependent term numerically becomes ~ 0.62 (the term 0.82 of (163) has been multiplied by $3/4$). Adding all the quantum contributions one has 0.89, which increases the earlier numerical value 0.86 of (163) in $\sim 3.5\%$. In fact, this is a small correction to the already known value which was obtained using the ansatz (62) in [20, 21].

7.3. Possible Vibrational Modes and the GSG Model

The only static soliton configurations with baryon number N_C, which emerge in the strong-coupling regime of QCD_2, are the ones we have considered above in eqs. (168)-(170). Precisely, these are the one-solitons of the GSG model which, in subsection 7.2., have been the subject of semi-classical treatment. Their quantum corrections by time-dependent rotations in flavor space have been computed, we focused on the one with the lowest classical mass. Since in two dimensions there are no spin degrees of freedom, in order to search for higher excitations we must look for vibrational modes which might in principle exist. These type of excitations in the strong coupling limit can be found as classical time-dependent solutions of the GSG equations of motion (11)-(12). Looking at time-dependent solutions of type (169) [see eq. (84)] one has that the field φ_2 satisfies ordinary sine-Gordon equation

$$\partial_{tt}\varphi_2 - \partial_{xx}\varphi_2 + 2\widetilde{m}^2\sqrt{\frac{4\pi}{N_C}}\sin\left(\frac{4\pi}{N_C}\varphi_2\right), \quad \varphi_1(x, t) \equiv 0. \quad (189)$$

The time dependent one-soliton solution of (189) for the field φ_2, determines the configuration $\{\Phi_1, \Phi_2, \Phi_3\}$ in (169) with baryon number N_C in the QCD_2 context. To look for higher excitations, for example, one can search for a coupled state of one-baryon and *breather* type vibrations (soliton-antisoliton bound states) of the GSG system, which can give a total baryon number N_C. We were not be able to find a more general time-dependent mixed single-baryon plus vibrational state with baryon number N_C for the general GSG equation. For example, this type of solution, if it exists, may be useful in order to study meson-baryon scattering as considered in [24]. As it is well known the SG eq. (189) does give vibrational solutions in the form of breather states (meson states), for later use we simply recall that in the large N_C limit the lowest-lying mesons have masses of order $\sqrt{m_q e_c}$ [46] (m_q is defined in eq. (194) below). We refer the reader to ref. [21] for more discussion, such as the various meson couplings to baryons with different degrees of exoticity.

8.   The GSG Solitons and the Exotic Baryons

8.1.   The First Exotic Baryon

Here we will follow the analog of the rigid-rotor approach (RRA) to quantize solitons and obtain exotic states. In this method it is assumed that the higher order representation multiplets are different rotational (in spin and isospin) states of the same object (the "classical baryon", i.e the soliton field) [31]. This assumption has allowed in the past the obtention of some relations between the characteristics of the nonexotic baryon multiplets which are satisfied up to a few percent in nature. However, see refs. [47, 48] for some critiques to this conventional approach for exotic baryons. According to these authors the conventional RRA, in which the collective rotational approach and vibrational modes of the soliton are assumed to be decoupled, and only the rotational modes are quantized, is only justified at large N_C for nonexotic collective states in $SU(3)$ models. On the other hand, the bound state approach (BSA) to quantize solitons, due to Callan-Klebanov [48], considers broken $SU(3)$ symmetry in which the excitations carrying strangeness are taken as vibrational modes, and should be quantized as harmonic vibrations. However, for exotic states the Callan-Klebanov approach does not reproduce the RRA result; indeed this approach gives no exotic resonant states when applied to the original Skyrme model [48]. There was intensive discussion of connections between the both approaches mentioned above. The rotation-vibration approach (RVA) (see [49] and references therein) includes both rotational (zero modes) and vibrational degrees of freedom of solitons and is a generalization of the both methods above, which therefore appear in some regions of the RVA method when certain degrees of freedom are frozen. A major result of the RVA method is that pentaquark states *do* indeed emerge in both methods above, i.e. in the RRA and BSA. In order to illustrate the present situation of the theoretical controversy let us mention that the RVA approach was criticized in [50], and the reply to this criticism was given in [51].

Following the analog of the RRA, the expression (186) can be used to compute the energy of the first exotic baryon $\mathcal{E}_1$ (a state containing $N_C + 1$ quarks and one antiquark) by taking the corresponding Casimir $C_2(\mathcal{E}_1)$ for $R = \mathbf{35}$ of flavor relevant to the exotic state in two-dimensions. This state is an analogue of the $\overline{\mathbf{10}}, \mathbf{27}$ and $\mathbf{35}$ states in four dimensions. So, following [21], in the conventional RRA one has that the mass of the first exotic state becomes

$$M(\mathcal{E}_1) = M(classical)\left\{1 + \frac{3}{4}\left[\frac{\pi^2}{8}\frac{1}{N_C}\left(3 + N_f - \frac{6}{N_f}\right) + \frac{3\pi^2}{8}\frac{1}{N_C^2}\left(N_f - \frac{3}{N_f}\right)\right]\right.$$
$$\left. - \frac{3\pi^2}{32}\frac{(N_f - 1)^2}{N_f} + \frac{3}{2}\right\}(190)$$

In the interesting case $N_C = 3$, $N_f = 3$ this becomes

$$M(\mathbf{35}) = M(classical)\left\{1 + \frac{3}{4}\frac{\pi^2}{4} - \frac{\pi^2}{8} + \frac{3}{2}\right\}. \tag{191}$$

In this case the correction due to quantum fluctuations around the classical solution is still larger than the classical term, as it was in the earlier computation (165). However, numerically in eq. (191) the correction is 2.12, whereas in eq. (165) it was 2.46. In fact, the contribution in (191) decreases in 0.34 units the earlier computation. So, we may claim that the introduction of unequal quark masses and the ansatz given by the GSG model slightly improve the semi-classical approximation.

Moreover, notice that the ratio of the experimental masses of the $\Theta^+(1530)$ and the nucleon is 1.63. On the other hand, the ratio of the first exotic to that of the lightest baryon in the QCD$_2$ model becomes

$$\frac{M_{35}}{M_{10}} = \frac{1 + \frac{3\pi^2}{16} - \frac{\pi^2}{8} + \frac{3}{2}}{1 + \frac{\pi^2}{16} - \frac{\pi^2}{8} + \frac{3}{2}} \sim 1.65, \tag{192}$$

which is only 1% larger to its 4D analog. This must be compared to the earlier calculation which gave a value 17% larger [see eq. (165)]. However, the result in (192) could be a numerical coincidence, since in two dimensions we are not considering the spin degrees of freedom that is important in QCD$_4$, even though the effects of unequal quark masses $m_3 < m_1 < m_2$ have been incorporated as an exact (without using perturbation theory) contribution to the hamiltonian.

8.2. Exotic Baryon Higher Multiplets

Let us consider exotic states $\mathcal{E}_p$ containing p antiquarks and $N_C + p$ quarks. In the case $N_C = 3$, $N_f = 3$, the only allowed $\mathcal{E}_2$ state is a $\mathbf{81}$ representation of flavor. In the particular case $N_f = 3$, for general N_C the mass of the $\mathcal{E}_p$ state is

$$M(\mathcal{E}_p) = M(classical)\left\{1 + \frac{3}{4}(\frac{\pi}{2N_C})^2\left[N_C(p+1) + p(p+2) - \frac{2}{3}N_C^2\right] + 3/2\right\}, \tag{193}$$

where the correction is considerably larger than unity. For example for $N_C = 3$ the mass correction becomes 3.76 units. Even though this correction is one unit less than the one obtained in [21], we would not consider it as a consistent semi-classical approximation for $N_C = 3$. However, we may consider the spacing Δ between $\mathcal{E}_{p+1}$ and $\mathcal{E}_p$ exotic states, which for large N_C becomes

$$\Delta \equiv \mathcal{E}_{p+1} - \mathcal{E}_p = (\frac{3}{4})\frac{\pi^2}{4}\frac{M_{classical}}{N_C} \sim 3.8\sqrt{e_c m_q}; \qquad m_q \equiv \frac{2m_1 + m_2}{3} \tag{194}$$

so, the constant Δ of [21] is decreased by a factor of 3/4. Since $M_{classical}$ is $\mathcal{O}(N_C^1)$, then the parameter Δ is a constant $\mathcal{O}(N_C^0)$ as the exoticity p is increased. Notice that the low-lying mesons masses are $\mathcal{O}(N_C^0)$ in the large N_C limit [21]. This would mean that the constant Δ value is like the addition of a meson to the $p-$state, in the form of quark-antiquark pair, in order to progress to the next excitation $p + 1$ [52]. Remember that the low-lying mesons in the SG theory have masses $\sim 3.2\sqrt{m_q e_c}$ [46], which are very close to the spacing Δ defined in (194).

8.3. Radius Parameter of the QCD$_2$ Exotic Baryons

In QCD$_2$, as found above, the quantum correction to the mass depends on one analogue of the moment of inertia appearing in four dimensions. Following [21] one considers

$$I = M(classical) < r^2 >, \tag{195}$$

the effective soliton radius can be defined by

$$<< r >> \equiv \sqrt{< r^2 >}. \tag{196}$$

Let us compare the quantum mass formula (193) with the corresponding relation in four dimensions [31] in the large N_C limit ($N_C >> p >> 1$), so one has

$$I = \frac{8N_C^2}{3\pi^2 M_{classical}}, \tag{197}$$

and then

$$<< r >> = \sqrt{\frac{I}{M_{classical}}} = \sqrt{\frac{8}{3}} \frac{N}{\pi M_{classical}} = \frac{1}{0.96\pi N_f^{1/4} \sqrt{e_c m_q}}, \tag{198}$$

where m_q was defined in (194). For $N_f = 3$ flavors, $e_c = 100 MeV$ for the coupling, and quark masses $m_3 = 4$ MeV, $m_1 = 54.5$ MeV, and $m_2 = 55.1$ MeV [these values satisfy the relationship $13m_3 = 5m_1 - 4m_2$ relevant in two-dimensions as is obtained from (86) and (115)], we get for the effective baryon radius $\approx 1/(294 MeV) \sim 0.7$ fm. This is 12.5% less than the radius estimated in [21] for QCD$_2$ exotic baryons. As a curiosity, notice that the radius parameter of Θ^+ has been estimated to be around 1.13 fm $= 5.65$ GeV^{-1} (see e.g. [53] and references therein).

9. Discussion

The generalized sine-Gordon model GSG (11)-(12) provides a variety of soliton and kink type solutions. The appearance of the non-integrable double sine-Gordon model as a sub-model of the GSG model suggests that this model is a non-integrable theory for the arbitrary set of values of the parameter space. However, a subset of values in parameter space determine some reduced sub-models which are integrable, e.g. the sine-Gordon submodels of subsections 5.1., 5.2. and 5.3..

In connection to the ATM spinors it was suggested that they are confined inside the GSG solitons and kinks since the gauge fixing procedure does not alter the $U(1)$ and topological currents equivalence (36). Then, in order to observe the bag model confinement mechanism it is not necessary to solve for the spinor fields since it naturally arises from the currents equivalence relation. In this way our model presents a bag model like confinement mechanism as is expected in QCD.

Besides, through the bosonization process it has been shown that the (generalized) massive Thirring model (GMT) corresponds to the GSG model [7], therefore, in view of the GSG solitons and kinks found above we expect that the spectrum of the GMT model will contain 4 solitons and their relevant anti-solitons, as well as the kink and antikink excitations. The GMT Lagrangian describes three flavor massive spinors with current-current interactions among themselves. So, the total number of solitons which appear in the bosonized sector suggests that the additional soliton (fermion) is formed due to the interactions between the currents in the GMT sector. However, in subsection 5.3. the soliton masses M_3 and M_4 become the same for the case $\mu_1 = \mu_2$, consequently, for this case we have just three solitons in the GSG spectrum, i.e., the ones with masses M_1, M_2 (subsections 5.1.-5.2.) and $M_3 = M_4$ (subsection 5.3.), which will correspond in this case to each fermion flavor of the GMT model. Moreover, the $sl(3,\mathbb{C})$ GSG model potential (7) has the same structure as the effective Lagrangian of the massive Schwinger model with $N_f = 3$ fermions, for a convenient value of the vacuum angle θ. The multiflavor Schwinger model resembles with four-dimensional QCD in many respects (see e.g. [54] and references therein).

In view of the discussions above the $sl(n,\mathbb{C})$ ATM models may be relevant in the construction of the low-energy effective theories of multiflavor QCD$_2$ with the dynamical fermions in the fundamental and adjoint representations. Notice that in the ATM models the Noether and topological currents and the generalized sine-Gordon/massive Thirring models equivalences take place at the classical [6, 29] and quantum mechanical level [7, 30].

On the other hand, the interest in baryons with exotic quantum numbers has recently been stimulated by various reports of baryons composed by four quarks and an antiquark. The existence of these baryons cannot yet be regarded as confirmed, however, reports of their existence have stimulated new investigations about baryon structure (see e.g. [55, 16] and references therein). Recently, the spectrum of exotic baryons in QCD$_2$, with $SU(N_f)$ flavor symmetry, has been discussed providing strong support to the chiral-soliton picture for the structure of normal and exotic baryons in four dimensions [21]. The new puzzles in non-perturbative QCD are related to systems with unequal quark masses, so the QCD$_2$ calculation must take into account the $SU(N_f)$-breaking mass effects, i.e. for $N_f = 3$ it must be $m_s \neq m_{u,d}$. We have extended the results of refs. [20, 21] concerning several properties of normal and exotic baryons by including unequal quark mass parameters. In the case of $N_f = 3$ flavors, the low-energy hadron states are described by the $su(3)$ generalized sine-Gordon model, providing a framework for the exact computations of the lowest-order quantum corrections of various quantities, such as the masses of the normal and exotic baryons. The semi-classical quantization method we adopted is an analogue of the rigid-rotor approach (RRA) applied in four dimensional QCD to quantize normal and exotic baryons (see e.g. [31]). Even though there is no spin in 2D, we have compared our results to their analogues in 4D; so, obtaining various similarities to the results from the chiral-soliton approaches in QCD$_4$. The RRA we have followed, as discussed in section 8., may be justified in our case since there is no mixing between the intrinsic vibrational modes and the collective rotation in flavor space degrees of freedom [47]. It is remarkable that the GSG ansatz (60), with soliton solutions which take into account the unequal quark mass parameters, allowed us to improve the lowest order quantum corrections for various physical quantities, such as the baryon masses; in this way rendering the semi-classical method more reliable in the large N_C limit. Other properties of the baryons such as a proper treatment of $k-$baryon bound states (extending the results of [42] for GSG type ansatz), including baryon-meson scattering amplitudes, are still to be addressed in the future.

Finally, we have found that the remarkable double sine Gordon model arises as a reduced GSG model bearing a kink(K) type solution describing a multi-baryon; so, the description of the multiflavor spectrum and some resonances in QCD $_2$ may take advantage of the properties of the DSG semiclassical spectrum and $K\bar{K}$ system which are being considered in the current literature [3, 4, 39, 40].

Acknowledgements

I would like to thank the Physics Department-UFMT (Cuiabá) for hospitality and CNPq-FAPEMAT for support. I am also very gratefully to H.L. Carrion for collaboration in a previous work and Dr. G. Takács and Prof. N. Ohta for communications and valuable comments about their earlier works.

References

[1] G.Delfino and G. Mussardo, *Nucl. Phys.* **B516** (1998) 6675.

[2] Z. Bajnok, L.Palla, G. Takács and F. Wágner, *Nucl. Phys.* **B601** (2001) 503.

[3] D. K. Campbell, M. Peyrard and P. Sodano, *Physica* **D19** (1986) 165.

[4] G. Mussardo, V. Riva and G. Sotkov, *Nucl. Phys.* **B687** (2004) 189.
     D. Controzzi and G. Mussardo, *Phys. Rev. Lett.* **92** (2004) 021601.

[5] J. Acosta, H. Blas, *J. Math. Phys.* **43** (2002) 1916,
     H.Blas, "Generalized sine-Gordon and massive Thirring models," in *New Developments in Soliton Research*, pp.123-147, Ed. L.V. Chen, Nova Science Pub. 2006; [arXiv:hep-th/0407020].

[6] H. Blas, *JHEP* **0311** (2003) 054, see also hep-th/0407020.

[7] H. Blas, *Eur. Phys. J.* **C37** (2004) 251;
     "Bosonization, soliton particle duality and Mandelstam-Halpern operators," in *Trends in Boson Research*, pp. 79-108, Ed. A.V. Ling ; Nova Science Pub. 2006;[arxiv:hep-th/0409269].

[8] H. Blas, H. L. Carrion and M. Rojas, *JHEP* **0503** (2005) 037;
     H. Blas, *JHEP* **0506** (2005) 022;
     H.Blas and H. L. Carrion, *Noncommutative solitons and kinks in affine Toda model coupled to matter and extended hadron model*, to appear.

[9] H. Blas and H.L. Carrion, *JHEP* **0701** (2007) 027; [hep-th/0610107].

[10] S.-J. Chang, S.D. Ellis and B.W. Lee, *Phys. Rev.* **D11** (1975) 3572.

[11] T. Uchiyama, *Phys. Rev.* **D14** (1976) 3520.

[12] B. Aubert *et al.* [BABAR Collaboration], *Phys. Rev. Lett.* **90** (2003) 242001;
V. V. Barmin *et al.* [DIANA Collaboration], *Phys. Atom. Nucl.* **66** (2003) 1715;
D. Besson *et al.* [CLEO Collaboration], *AIP Conf. Proc.* **698** (2004) 497;
M. Ablikim *et al.* [BES Collaboration], *Phys. Rev. Lett.* **93** (2004) 112002.

[13] R. A. Schumacher, *AIP Conf. Proc.* **842** (2006) 409; nucl-ex/0512042.
B. McKinnon, K. Hicks et al. (CLAS Collaboration), hep-ex/0603028.
V.V. Barmin et al. (DIANA Collaboration), hep-ex/0603017.

[14] V. V. Barmin et al. [DIANA Collaboration], hep-ex/0603017;
A. Kubarovsky, V. Popov and V. Volkov [for the SVD-2 Collaboration], hep-ex/0610050.

[15] B. McKinnon et al. [CLAS Collaboration], *Phys. Rev. Lett.* **96** (2006) 212001.

[16] D. Diakonov,*AIP Conf. Proc.* **892** (2007) 258; [arxiv.org:hep-ph/0610166].
Y. Azimov, K. Goeke and I. Strakovsky, *Phys. Rev.* **D76** (2007) 074013; [arXiv:hep-ph/07082675].

[17] E. Abdalla, M.C.B. Abdalla and K.D. Rothe, Non-perturvative methods in two-dimensional quantum field theory, 2nd edition (World Scientific, Singapore, 2001).

[18] D.J. Gross, I.R. Klebanov, A.V. Matytsin and A.V. Smilga; *Nucl. Phys.* **B461** (1996) 109.

[19] G. t Hooft, *Nucl. Phys.* **B75** (1974) 461.

[20] Y. Frishman and J. Sonnenschein, *Phys. Reports* **223** (1993) 309.

[21] J. R. Ellis and Y. Frishman, *JHEP* **0508** (2005) 081.

[22] H. Blas, *JHEP* **0703** (2007) 055; [arXiv:hep-th/0702197].

[23] J. R. Ellis, Y. Frishman, A. Hanany and M. Karliner, *Nucl. Phys.* **B382** (1992) 189.

[24] J. R. Ellis, Y. Frishman and M. Karliner, *Phys. Lett.* **566B** (2003) 201;
Y. Frishman and M. Karliner, *Phys. Lett.* **541B** (2002) 273.

[25] L.A. Ferreira, J-L. Gervais, J. Sánchez Guillen and M.V. Saveliev, *Nucl. Phys.* **B470** (1996) 236.

[26] H. Blas, *Phys. Rev.* **D66** (2002) 127701.

[27] A. Armoni, Y. Frishman, J. Sonnenschein, *Phys. Rev. Lett.* **80** (1998) 430; *Int. J. Mod. Phys.* **A14** (1999) 2475.

[28] H. Blas, *Nucl. Phys.* **B596** (2001) 471; see also [arXiv:hep-th/0005037].

[29] H. Blas and B.M. Pimentel, *Annals Phys.* **282** (2000) 67.
H. Blas, *Nucl. Phys.* **B596** (2001) 471; see also [arXiv:hep-th/0005037].

[30] H. Blas and L.A. Ferreira, *Nucl. Phys.* **B571** (2000) 607.

[31] D. Diakonov, V. Petrov and M. V. Polyakov, *Z.Phys.* **A359** (1997) 305.

[32] Y.S. Kivshar and B.A. Malomed, *Rev. Mod. Phys.* **61** (1989) 763.

[33] C.T. Zhang, *Phys. Rev.* **A35** (1987) 886.
B.A. Malomed, *Phys. Lett.* **123A** (1987) 459.

[34] A.G. Bueno, L.A. Ferreira and A.V. Razumov, *Nucl. Phys.* **B626** (2002) 463.

[35] J.M. Humphreys, Introduction to Lie algebras and representation theory, Graduate Texts in Mathematics, Vol. 9, Springer-Verlag, 1972.

[36] D. Diakonov and V. Y. Petrov, *Nucl. Phys.* **B272** (1986) 457;
D. Diakonov, "From pions to pentaquarks," arXiv:hep-ph/0406043.

[37] Y. Frishman and J. Sonnenschein, *Nucl. Phys.* **B496** (1997) 285.

[38] E. Witten, *Annals Phys.* **128** (1980) 363.
N. Ohta, *Prog. Theor. Phys.* **66** (1981) 1408; [*Erratum-ibid.* **A67** (1982) 993].

[39] G. Mussardo, *Nucl. Phys.* **B779** (2007) 101.

[40] G. Takacs and F. Wagner, *Nucl. Phys.* **B741** (2006) 353.

[41] N. Riazi, A. Azizi and S. M. Zebarjad, *Phys. Rev.* **D66** (2002) 065003.

[42] Y. Frishman and W.J. Zakrzewski, *Nucl. Phys.* **B331** (1990) 781.

[43] S. Bertini, S.L. Cacciatori and B.L. Cerchiai, *J. Math. Phys.* **47** (2006) 043510;
[arxiv.org:math-ph/0510075].

[44] G.D. Date, Y. Frishman and J. Sonnenschein, *Nucl. Phys.* **B283** (1987) 365.

[45] E. Ravinovici, A. Schwimmer and S. Yankielowicz, *Nucl. Phys.* **B248** (1984) 523.

[46] R. Rajaraman, Solitons and Instantons, (1982), Elsevier, Amsterdam.

[47] N. Itzhaki, I. R. Klebanov, P. Ouyang and L. Rastelli, *Nucl. Phys.* **B684** (2004) 264;
T.D. Cohen, *Phys. Lett.* **581B** (2004) 175;
T.D. Cohen and R.F. Lebed, *Phys. Lett.* **578B** (2004) 150.

[48] C.G. Callan Jr. and I.R. Klebanov, *Nucl. Phys.* **B262** (1985) 365.

[49] H. Walliser and H. Weigel, *Eur. Phys. J.* **C26** (2005) 361.

[50] T. D. Cohen, hep-ph/0511174.

[51] H. Walliser and H. Weigel, hep-ph/0511297.

[52] D. Diakonov and V. Y. Petrov, *Phys. Rev.* **D69** (2004) 056002.

[53] A.Bhattacharya, B.Chakrabarti, S.Mani and S.N. Banerjee, On Some Properties of Exotic Baryons in Quasi particle diquark Model, [arXiv:hep-ph/0611367].

[54] Y. Hosotani and R. Rodriguez, *J. Physics* **A31** (1998) 9925.
J. E. Hetrick, Y. Hosotani and S. Iso, *Phys. Lett.* **350B** (1995) 92.

[55] S. Kabana, *J. Phys.* **G31** (2005) S1155.

In: High Energy Physics Research Advances
Editors: T.P. Harrison et al, pp. 81-94

ISBN 978-1-60456-304-7
© 2008 Nova Science Publishers, Inc.

Chapter 3

INTERMEDIATE ENERGY SPECTRUM OF FIVE COLOUR QCD AT ONE LOOP

M.L. Walker

Institute of Quantum Science, Nihon University,
Chiyoda, Japan 101-8308

Abstract

I consider the monopole condensate of five colour QCD. The näive lowest energy state is unobtainable at one-loop for five or more colours due to simple geometry. The consequent adjustment of the vacuum condensate generates a hierarchy of confinement scales in a natural Higgs-free manner. QCD and QED-like forces emerge naturally, acting upon matter fields that may be interpreted as down quarks, up quarks and electrons.

1. Introduction

It is already known [1, 2, 3], that $SU(N)$ QCD can lower the energy of its vacuum with a monopole background field along the Abelian directions, where the Abelian components are equal in magnitude but orthogonal in real space [2, 3]. This orthogonality, while of no special consequence in $SU(3)$ QCD in three space dimensions, does have consequences when the number of Abelian directions is greater than three. As noted originally by Flyvbjerg [2], $SU(N \geq 5)$ QCD cannot realise its true minimum because four orthogonal vectors cannot fit in three dimensions. I shall call a system kept from reaching its true lowest energy state by a lack of spatial dimensions *dimensionally frustrated*.

The research in this chapter seeks to identify the monopole condensate of five-colour QCD, or at least a good candidate for it, and examine the consequences. It assumes the dual superconductor model of confinement [4, 5, 6, 7, 8] in which chromomagnetic monopole-antimonopole pairs play the dual role to Cooper pairs, restricting the electric component of the chromodynamic field to flux tubes. This model is by no means proven but the case for it is very strong. I shall handle the monopole degrees of freedom with the Cho-Faddeev-Niemi-Shabanov decomposition [7, 9, 10, 11] which specifies the internal directions corresponding to the Abelian generators in a gauge-covariant way, automatically introducing the monopole field in the process. It is explained in detail in section 2..

At extremely high energies where the effects of confinement are not significant, the dynamics are simply those of $SU(5)$ QCD in the far ultraviolet. I shall show however that the dimensionally frustrated condensate is anisotropic and this causes some colours to be confined more tightly than others. Even more interesting, white combinations do not necessarily need to contain all five colours as one would expect. The first three colours, labelled red, blue and green, form unconfined combinations among themselves, as do the additional two colours, which I have called ultraviolet and infrared. (Actually the confining effect is not quite zero but is much smaller than other effects and is therefore ignored.) This is but one way in which the $SU(3)$ symmetry naturally breaks off from the rest of the symmetry group.

Examination of the gluon dynamics in section 4. finds that those corresponding to one particular root vector are confined more tightly than the rest. At intermediate energies these gluons drop out of the dynamics, causing the coupling constants of those that remain to scale differently and again leading to the separation of $SU(3)$. The emergence of an unconfined Abelian gauge field that can be identified with the photon is demonstrated in section 5.. While these sections are chiefly reviews of work already completed [12], section 6. contains new material concerning the matter field representations and identifies the neutral, or white, colour combinations as well as the natural emergence of both up and down quarks and the electron, all with the correct relative electric and colour charges.

While this is strictly speaking an examination of a one-loop effect in $SU(5)$ QCD, the prospect of grand unification does arise. The conclusion of this chapter is that its effective theory does not include weak nuclear decay but that it could well describe a unification of QCD with QED. The value of such a unification is also discussed along with the predictions of this theory in section 7., before ending with a summary in section 8..

2.   The Cho-Faddeev-Niemi-Shabanov Decomposition

My treatment of the monopole condensate rests on the Cho-Faddeev-Niemi-Shabanov decomposition [7, 9, 10, 11]. I use the following notation:
The Lie group $SU(N)$ has $N^2 - 1$ generators $\lambda^{(j)}$, of which $N - 1$ are Abelian generators $\Lambda^{(i)}$. For simplicity, I specify the gauge transformed Abelian directions (Cartan generators) with

$$\hat{\mathbf{n}}_i = U^\dagger \Lambda^{(i)} U. \tag{1}$$

In the same way, I replace the standard raising and lowering operators $E_{\pm\alpha}$ for the root vectors α with the gauge transformed ones

$$E_{\pm\alpha} \to U^\dagger E_{\pm\alpha} U, \tag{2}$$

where $E_{\pm\alpha}$ refers to the gauge transformed operator throughout the rest of this chapter.

Gluon fluctuations in the $\hat{\mathbf{n}}_i$ directions are described by $c_\mu^{(i)}$. The gauge field of the covariant derivative which leaves the $\hat{\mathbf{n}}_i$ invariant is

$$g\mathbf{V}_\mu \times \hat{\mathbf{n}}_i = -\partial_\mu \hat{\mathbf{n}}_i. \tag{3}$$

In general this is

$$\mathbf{V}_\mu = c_\mu^{(i)} \hat{\mathbf{n}}_i + \mathbf{B}_\mu, \quad \mathbf{B}_\mu = g^{-1} \partial_\mu \hat{\mathbf{n}}_i \times \hat{\mathbf{n}}_i, \tag{4}$$

where summation is implied over i. $\mathbf{B}_\mu$ can be a attributed to non-Abelian monopoles, as indicated by the $\hat{n}_i$ describing the homotopy group $\pi_2[SU(N)/U(1)^{\otimes(N-1)}] \approx \pi_1[U(1)^{\otimes(N-1)}]$. The monopole field strength

$$\mathbf{H}_{\mu\nu} = \partial_\mu \mathbf{B}_\nu - \partial_\nu \mathbf{B}_\mu + g\mathbf{B}_\mu \times \mathbf{B}_\nu, \tag{5}$$

has only Abelian components, *ie.*

$$H_{\mu\nu}^{(i)} \hat{n}_i = \mathbf{H}_{\mu\nu}, \tag{6}$$

where $H_{\mu\nu}^{(i)}$ has the eigenvalue $H^{(i)}$. Since I am only concerned with magnetic backgrounds, $H^{(i)}$ is considered the magnitude of a background magnetic field $\mathbf{H}^{(i)}$. The field strength of the Abelian components $c_\mu^{(i)}$ also lies in the Abelian directions as expected and is shown by

$$\mathbf{F}_{\mu\nu} = F_{\mu\nu}^{(i)} \hat{n}_i, \tag{7}$$

where

$$F_{\mu\nu}^{(i)} = \partial_\mu c_\nu^{(i)} - \partial_\nu c_\mu^{(i)}. \tag{8}$$

The Lagrangian of the Abelian and monopole components is

$$-\frac{1}{4}(F_{\mu\nu}^{(i)} \hat{n}_i + \mathbf{H}_{\mu\nu})^2 \tag{9}$$

The dynamical degrees of freedom (DOF) perpendicular to $\hat{n}_i$ are denoted by $\mathbf{X}_\mu$, so if $\mathbf{A}_\mu$ is the gluon field then

$$\mathbf{A}_\mu = \mathbf{V}_\mu + \mathbf{X}_\mu = c_\mu^{(i)} \hat{n}_i + \mathbf{B}_\mu + \mathbf{X}_\mu, \tag{10}$$

where

$$\mathbf{X}_\mu \perp \hat{n}_i, \quad \mathbf{X}_\mu = g^{-1} \hat{n}_i \times \mathbf{D}_\mu \hat{n}_i, \quad \mathbf{D}_\mu = \partial_\mu + g\mathbf{A}_\mu \times . \tag{11}$$

Because $\mathbf{X}_\mu$ is orthogonal to all Abelian directions it can be expressed as a linear combination of the raising and lowering operators $E_{\pm\alpha}$, which leads to the definition

$$X_\mu^{(\pm\alpha)} \equiv E_{\pm\alpha} \mathrm{Tr}[\mathbf{X}_\mu E_{\pm\alpha}], \tag{12}$$

so

$$X_\mu^{(-\alpha)} = X_\mu^{(+\alpha)\dagger}. \tag{13}$$

$\mathbf{H}_{\mu\nu}^{(\alpha)}$, defined by

$$\mathbf{H}_{\mu\nu}^{(\alpha)} = \alpha_j H_{\mu\nu}^{(j)}, \tag{14}$$

is the monopole field strength tensor felt by $\mathbf{X}_\mu^{(\alpha)}$. I also define the background magnetic field

$$\mathbf{H}^{(\alpha)} = \alpha_j \mathbf{H}^{(j)}, \tag{15}$$

whose magnitude $H^{(\alpha)}$ is $\mathbf{H}_{\mu\nu}^{(\alpha)}$'s non-zero eigenvalue. Since both $\mathbf{B}_\mu, \mathbf{X}_\mu$ contain off-diagonal degrees of freedom, it is worth clarifying that $\mathbf{X}_\mu$ contains the quantum fluctuations taking place on a generally non-trivial background whose topology is contained in the monopole field $\mathbf{B}_\mu$.

3.   The Vacuum State of Five-Colour QCD

The one-loop effective energy of five-colour QCD is given by [2, 3]

$$\mathcal{H} = \sum_{\alpha > 0} \|\mathbf{H}^{(\alpha)}\|^2 \left[\frac{1}{5g^2} + \frac{11}{48\pi^2} \ln \frac{H^{(\alpha)}}{\mu^2} \right] \tag{16}$$

which is minimal when

$$H^{(\alpha)} = \mu^2 \exp\left(-\frac{1}{2} - \frac{48\pi^2}{55g^2} \right). \tag{17}$$

This neglects an alleged imaginary component [13] which has been called into serious question recently [14, 15, 16, 17, 18, 19, 3] with more and more studies finding that it is only an artifact of the quadratic approximation. Taking this to be the case, I employ the Savvidy vacuum. This can be criticised for lacking Lorentz covariance but I argue that it is likely to match the true vacuum at least locally.

Since

$$\|\mathbf{H}^{(1,0,0,0)}\| = \|\mathbf{H}^{(1)}\|,$$

$$\left\|\mathbf{H}^{\left(\pm\frac{1}{2}, \frac{\sqrt{3}}{2}, 0, 0\right)}\right\|^2 = \frac{1}{4}\|\mathbf{H}^{(1)}\|^2 + \frac{3}{4}\|\mathbf{H}^{(2)}\|^2 \pm \frac{\sqrt{3}}{2}\mathbf{H}^{(1)} \cdot \mathbf{H}^{(2)},$$

$$\left\|\mathbf{H}^{\left(\pm\frac{1}{2}, \frac{1}{\sqrt{12}}, \frac{2}{\sqrt{6}}, 0\right)}\right\|^2 = \frac{1}{4}\|\mathbf{H}^{(1)}\|^2 + \frac{1}{12}\|\mathbf{H}^{(2)}\|^2 + \frac{2}{3}\|\mathbf{H}^{(3)}\|^2 \pm \sqrt{\frac{2}{3}}\mathbf{H}^{(1)} \cdot \mathbf{H}^{(2)}$$
$$\pm \frac{1}{2\sqrt{3}}\mathbf{H}^{(1)} \cdot \mathbf{H}^{(3)} + \frac{\sqrt{2}}{3}\mathbf{H}^{(2)} \cdot \mathbf{H}^{(3)},$$

$$\left\|\mathbf{H}^{\left(0, -\frac{1}{\sqrt{3}}, \frac{2}{\sqrt{6}}, 0\right)}\right\|^2 = \frac{1}{3}\|\mathbf{H}^{(2)}\|^2 + \frac{2}{3}\|\mathbf{H}^{(3)}\|^2 - \frac{2\sqrt{2}}{3}\mathbf{H}^{(2)} \cdot \mathbf{H}^{(3)},$$

$$\left\|\mathbf{H}^{\left(0, 0, -\frac{\sqrt{3}}{\sqrt{8}}, \frac{\sqrt{5}}{\sqrt{8}}\right)}\right\|^2 = \frac{3}{8}\|\mathbf{H}^{(3)}\|^2 + \frac{5}{8}\|\mathbf{H}^{(4)}\|^2 - \frac{\sqrt{15}}{4}\mathbf{H}^{(3)} \cdot \mathbf{H}^{(4)},$$

$$\left\|\mathbf{H}^{\left(0, -\frac{\sqrt{3}}{\sqrt{8}}, \frac{1}{\sqrt{24}}, \frac{\sqrt{5}}{\sqrt{8}}\right)}\right\|^2 = \frac{3}{8}\|\mathbf{H}^{(2)}\|^2 + \frac{1}{24}\|\mathbf{H}^{(3)}\|^2 + \frac{\sqrt{5}}{\sqrt{8}}\|\mathbf{H}^{(4)}\|^2 - \sqrt{\frac{1}{16}}\mathbf{H}^{(2)} \cdot \mathbf{H}^{(3)}$$
$$- \frac{\sqrt{15}}{4}\mathbf{H}^{(2)} \cdot \mathbf{H}^{(4)} + \frac{\sqrt{5}}{\sqrt{48}}\mathbf{H}^{(3)} \cdot \mathbf{H}^{(4)},$$

$$\left\|\mathbf{H}^{\left(\pm\frac{1}{2}, \frac{1}{\sqrt{12}}, \frac{1}{\sqrt{24}}, \frac{\sqrt{5}}{\sqrt{8}}\right)}\right\|^2 = \frac{1}{4}\|\mathbf{H}^{(1)}\|^2 + \frac{1}{12}\|\mathbf{H}^{(2)}\|^2 + \frac{1}{24}\|\mathbf{H}^{(3)}\|^2 + \frac{\sqrt{5}}{\sqrt{8}}\|\mathbf{H}^{(4)}\|^2$$
$$\pm \sqrt{\frac{2}{3}}\mathbf{H}^{(1)} \cdot \mathbf{H}^{(2)} \pm \frac{1}{\sqrt{24}}\mathbf{H}^{(1)} \cdot \mathbf{H}^{(3)} + \frac{1}{\sqrt{16}}\mathbf{H}^{(2)} \cdot \mathbf{H}^{(3)}$$
$$\pm \frac{\sqrt{5}}{\sqrt{8}}\mathbf{H}^{(1)} \cdot \mathbf{H}^{(4)} + \frac{\sqrt{5}}{\sqrt{24}}\mathbf{H}^{(2)} \cdot \mathbf{H}^{(4)} + \frac{\sqrt{5}}{\sqrt{48}}\mathbf{H}^{(3)} \cdot \mathbf{H}^{(4)} \tag{18}$$

it follows that

$$\|\mathbf{H}^{(i)}\| = \|\mathbf{H}^{(j)}\|, \quad \mathbf{H}^{(i)} \perp \mathbf{H}^{(j)}, \quad i \neq j, \tag{19}$$

Table 1. Candidate parallel components for vacuum condensate. The column on the left is for parallel vectors, the column on the right is for antiparallel vectors. $\Delta\mathcal{H}$ should be multiplied by $H^2\frac{11}{96\pi^2}$.

$\mathbf{H}^{(i)} = +\mathbf{H}^{(j)}$	$\Delta\mathcal{H}$	$\mathbf{H}^{(i)} = -\mathbf{H}^{(j)}$	$\Delta\mathcal{H}$
$\mathbf{H}^{(1)} = +\mathbf{H}^{(2)}$	1.06381	$\mathbf{H}^{(1)} = -\mathbf{H}^{(2)}$	1.06381
$\mathbf{H}^{(1)} = +\mathbf{H}^{(3)}$	0.857072	$\mathbf{H}^{(1)} = -\mathbf{H}^{(3)}$	0.857072
$\mathbf{H}^{(1)} = +\mathbf{H}^{(4)}$	0.715651	$\mathbf{H}^{(1)} = -\mathbf{H}^{(4)}$	0.715651
$\mathbf{H}^{(2)} = +\mathbf{H}^{(3)}$	1.01655	$\mathbf{H}^{(2)} = -\mathbf{H}^{(3)}$	0.656584
$\mathbf{H}^{(2)} = +\mathbf{H}^{(4)}$	0.882589	$\mathbf{H}^{(2)} = -\mathbf{H}^{(4)}$	0.577976
$\mathbf{H}^{(3)} = +\mathbf{H}^{(4)}$	1.00042	$\mathbf{H}^{(3)} = -\mathbf{H}^{(4)}$	0.540983

which means that the chromomagnetic field components must be equal in magnitude but mutually orthogonal in the lowest energy state. However three dimensional space can only accomodate three mutually orthogonal vectors. Since the number of Cartan components, *ie.* components corresponding to Abelian generators, is always $N - 1$ in $SU(N)$ it follows that QCD with more than four colours cannot achieve such an arrangement.

One could substitute the Cartan basis $\mathbf{H}^{(i)}$ but this leads to intractable equations that cannot be solved analytically. It is reasonable to expect that the lowest attainable energy state is only slightly different from (16) and that this difference is due to the failure of mutual orthogonality. I therefore propose the ansatz that all Cartan components are equal in magnitude to what they would be in the absence of dimensional frustration, and that their relative orientations in real space are chosen so as to minimise the energy. In practice this means that three of the four are mutually orthogonal and the remaining one is a linear combination of those three. This remainder will increase the effective energy through its scalar products with the mutually orthogonal vectors but not all scalar products contribute equally, which follows from the form of the root vectors in eq. (18). This means that the orientation of the remaining real space vector in relation to the mutually orthogonal ones impacts the effective energy.

A little thought reveals that the lowest energy state should have only one scalar product contribute to it. The problem of finding the lowest available energy state therefore reduces to finding the scalar product that contributes to it the least. The six candidates are

$$\mathbf{H}^{(1)} \cdot \mathbf{H}^{(2)}, \mathbf{H}^{(1)} \cdot \mathbf{H}^{(3)}, \mathbf{H}^{(1)} \cdot \mathbf{H}^{(4)},$$
$$\mathbf{H}^{(2)} \cdot \mathbf{H}^{(3)}, \mathbf{H}^{(2)} \cdot \mathbf{H}^{(4)}, \mathbf{H}^{(3)} \cdot \mathbf{H}^{(4)}. \tag{20}$$

As can be seen from table 1, $\mathbf{H}^{(3)} = -\mathbf{H}^{(4)}$ (antiparallel) yields the lowest effective energy when all other scalar products are zero.

Substituting this result into (18) finds that all $\mathbf{H}^{(\alpha)}$ have the same magnitude except for those that couple to $\mathbf{H}^{(4)}$, namely $\mathbf{H}^{\left(?,?,?,\sqrt{\frac{5}{8}}\right)}$, where ? indicates that there are several possible values. The other background field strengths are

$$\|\mathbf{H}^{(\alpha)}\|^2 = H^2, \tag{21}$$

while the strongest is

$$\|\mathbf{H}^{\left(0,0,-\sqrt{\frac{3}{8}},\sqrt{\frac{5}{8}}\right)}\|^2 = H^2\left(1 + \frac{\sqrt{15}}{4}\right), \tag{22}$$

and the weakest are

$$\|\mathbf{H}^{\left(?,?,\frac{1}{\sqrt{24}},\sqrt{\frac{5}{8}}\right)}\|^2 = H^2\left(1 - \sqrt{\frac{5}{48}}\right). \tag{23}$$

Remember that the negative signs are affected by $\mathbf{H}^{(3)}$, $\mathbf{H}^{(4)}$ being antiparallel.

Assuming the dual superconductor model of confinement [4, 5, 6, 7, 8], it follows that different valence gluons and even different quarks (in the fundamental representation) will be confined with different strengths and therefore at different length scales. Those that feel the background $H^{\left(0,0,-\frac{\sqrt{3}}{\sqrt{8}},\frac{\sqrt{5}}{\sqrt{8}}\right)}$ will be confined the most strongly, those that feel the backgrounds of the form $H^{\left(?,?,\frac{1}{\sqrt{24}},\frac{\sqrt{5}}{\sqrt{8}}\right)}$ will be confined least strongly. The remainder will be confined with intermediate strength.

At highest energy then, we have the full dynamics of $SU(5)$ QCD. Moving down to some intermediate energy however, finds that the dynamics associated with the root vector $\left(0,0,-\frac{\sqrt{3}}{\sqrt{8}},\frac{\sqrt{5}}{\sqrt{8}}\right)$ are confined out of the dynamics. The remaining gluons interact among themselves. Moving to lower energy scales I find that those dynamics are all removed in their turn except for those corresponding to the root vectors $\left(?,?,\frac{1}{\sqrt{24}},\frac{\sqrt{5}}{\sqrt{8}}\right)$, almost leaving an $SU(2)$ gauge field interaction. I say 'almost' because I shall later demonstrate that the form of the monopole condensate is sufficiently different from the $SU(2)$ condensate to alter the dynamics, producing three confined $U(1)$ gauge fields, two of which are contained within $SU(3)$, a further unconfined $U(1)$ gauge field that may be identified with the photon, and three copies of the valence gluons of $SU(2)$. At lowest energies only the unconfined gauge field remains. In this way a hierarchy of confinement scales and effective dynamics emerges naturally, without the introduction of any *ad. hoc.* mechanisms like the Higgs field.

4. Intermediate Energy Dynamics

In constructing the hierarchical picture above, I began with $SU(5)$ and finished with $U(1)$ but had no apparent gauge group governing the dynamics at the intermediate energy scale. The dynamics of this energy scale will prove to be quite interesting.

To facilitate the discussion I introduce a notation inspired by the Dynkin diagram of $SU(5)$. The root vectors implicitly specified in eq. (18) are all linear combinations of a few basis vectors, which according to Lie algebra representation theory can be chosen for convenience. I take the basis vectors

$$(1,0,0,0), \left(-\frac{1}{2}, \frac{\sqrt{3}}{2}, 0, 0\right), \left(0, -\frac{1}{\sqrt{3}}, \sqrt{\frac{2}{3}}, 0\right), \left(0, 0, -\sqrt{\frac{3}{8}}, \sqrt{\frac{5}{8}}\right), \tag{24}$$

which I shall each represent by

$$\text{OXXX, XOXX, XXOX, XXXO,} \tag{25}$$

respectively. The remaining root vectors are sums of these basis vectors. In this notation their representation contains an 'O' if the corresponding basis vector is included and 'X' if it is not. For example the root vector

$$\left(\frac{1}{2}, \frac{\sqrt{3}}{2}, 0, 0\right) = (1, 0, 0, 0) + \left(-\frac{1}{2}, \frac{\sqrt{3}}{2}, 0, 0\right), \qquad (26)$$

is represented by

$$OOXX = OXXX + XOXX. \qquad (27)$$

When convenient, a '?' is used to indicate that either 'O' or 'X' might be substituted.

In addition to its brevity, this notation has the nice feature of making obvious which root vectors can be combined to form other root vectors because there are no root vectors containing 'X' with 'O's on either side. There is no OXXO for example.

The confinement of $X_\mu^{\left(0,0,-\frac{\sqrt{3}}{\sqrt{8}},\frac{\sqrt{5}}{\sqrt{8}}\right)}$, the valence gluon corresponding to XXXO, out of the dynamics directly affects only those remaining valence gluons that couple to it, those of root vectors of the form ??O?. The remaining gluons, corresponding to the root vectors OOXX, XOXX and OXXX (collectively given by ??XX), may still undergo the full set of interactions available to them at higher energies. It is easy to see that these are the root vectors that comprise the group $SU(3)$, to which the other valence gluons couple forming two six dimensional representations. Subsequent discussion shall extend the X,O,? notation to include the valence gluons corresponding to a root vector. Whether it is the gluon or the root vector that is meant will be clear from context.

Consider the beta function, or to be less imprecise, the scaling of the various gluon couplings. I shall now demonstrate that the loss to confinement of the root vector XXXO causes unequal corrections to the running of the couplings for different gluons. Since this is only an introductory paper the following analysis is only performed to one-loop.

The gluons ??XX, corresponding to the above-mentioned $SU(3)$, retain their original set of interactions. Performing the standard perturbative calculation [2] therefore yields the standard result for $SU(5)$ QCD. The remaining gluons do not. The absence of the maximally confined XXXO restricts their three-point vertices to those of $SU(4)$, since all root vectors are now of the form ???X. The same is not true of the four-point interactions, but the exceptions do not contribute to the scaling of the coupling constant at one-loop [20]. We have then that the $SU(3)$ subgroup's coupling scales differently from the rest of the unconfined gluons when the maximally confined valence gluons XXXO drop out.

The beta function is proportional to the number of colours in pure QCD at one-loop, so as the length scale increases, the coupling among gluons within the $SU(3)$ subgroup initially grows faster than the couplings involving the other gluons. As noted above, the $SU(3)$ couplings will initially scale as in the five-colour theory, while the remainder scale as though there were only four colours. This specific behaviour must soon change due to both non-perturbative contributions and because the non-$SU(3)$ gluons have a weaker coupling. A detailed understanding requires a nonperturbative analysis well beyond the scope of this chapter. Indeed, the application of one-loop perturbation theory at anything other than the far ultraviolet is questionable in itself. The point remains that the $SU(3)$ subgroup ??XX separates from the remaining gluons by its stronger coupling strength.

The symmetry reduction that takes place in this model is suggestive of boson mass generation but there appears to be no obvious specific mechanism. Kondo *et. al.* have argued for the spontaneous generation of mass through various non-trivial mechanisms [18, 21, 3]. This is consistent with the well-studied correlation between confinement and chiral symmetry breaking (see [22, 23, 24, 25] and references therein).

5. The Emergence of QED

Neglecting off-diagonal gluons, the equality $\mathbf{H}^{(3)} = -\mathbf{H}^{(4)}$ allows the change in variables

$$c_\mu^{(3)}\hat{\mathbf{n}}_3 \rightarrow \frac{1}{2}(c_\mu^{(3)}\hat{\mathbf{n}}_3 + c_\mu^{(4)}\hat{\mathbf{n}}_4) + \frac{1}{2}(c_\mu^{(3)}\hat{\mathbf{n}}_3 - c_\mu^{(4)}\hat{\mathbf{n}}_4) = \frac{1}{\sqrt{2}}(A_\mu + E_\mu),$$

$$c_\mu^{(4)}\hat{\mathbf{n}}_4 \rightarrow \frac{1}{2}(c_\mu^{(3)}\hat{\mathbf{n}}_3 + c_\mu^{(4)}\hat{\mathbf{n}}_4) - \frac{1}{2}(c_\mu^{(3)}\hat{\mathbf{n}}_3 - c_\mu^{(4)}\hat{\mathbf{n}}_4) = \frac{1}{\sqrt{2}}(A_\mu - E_\mu). \tag{28}$$

Substituting eqs (28) into the Abelian dynamics (9) finds that the antisymmetric combination E_μ couples to the background

$$H(\hat{\mathbf{n}}_3 - \hat{\mathbf{n}}_4),$$

but the symmetric combination A_μ does not. Again by the dual superconductor model the former is confined (along with $c_\mu^{(1)}\hat{\mathbf{n}}_1$, $c_\mu^{(2)}\hat{\mathbf{n}}_2$) while the latter is not. Since the electromagnetic field is long range it is natural to interpret A_μ as the photon.

6. Matter Field Representations

I have shown how the coupling of the gluons to the monopole background determines their confinement strength and subsequent phenomenology. I now consider the matter fields and focus in particular on the fundamental representation of $SU(5)$. The confinement of the fundamental representation is determined by the maximal stability group [26, 27], which for $SU(5)$ is $U(4) \approx SU(4) \otimes U(1)$, where for any given element $(\cdots \psi \cdots)^T$, the $SU(4)$ acts only on the remaining orthogonal elements while the $U(1)$ causes it inconsequential phase changes. This latter $U(1)$ describes the monopole condensate contributing to the confinement and is given by the corresponding weight of the fundamental representation. As a concrete example, consider the fundamental element $(0\,0\,0\,0\,\psi)^T$. The $SU(4)$ of its maximal stability group are the matrices

$$\begin{pmatrix} & & & 0 \\ & T_i & & 0 \\ & & & 0 \\ & & & 0 \\ 0\,0\,0\,0 & & 1 \end{pmatrix}, \tag{29}$$

where T_i are the standard $SU(4)$ matrices $T_1 \ldots T_{15}$, while the $U(1)$ is generated by $T_{24} = \frac{1}{\sqrt{20}} diag(1\,1\,1\,1\,-4)$. Therefore the only component of the chromomonopole condensate that contributes to the confinement of $(0\,0\,0\,0\,\psi)^T$ is that generated by T_{24}. This is what would be expected based on the weights of the fundamental representation, and indeed the main result of [27] is that the weight of the representation determines which components of the monopole condensate contribute to a given particle's confinement.

The weights of the fundamental representation of $SU(5)$ are

$$
\begin{aligned}
(1\,0\,0\,0\,0)^T &: \left(\frac{1}{2}, \frac{1}{\sqrt{12}}, \frac{1}{\sqrt{24}}, \frac{1}{\sqrt{40}}\right) \\
(0\,1\,0\,0\,0)^T &: \left(-\frac{1}{2}, \frac{1}{\sqrt{12}}, \frac{1}{\sqrt{24}}, \frac{1}{\sqrt{40}}\right) \\
(0\,0\,1\,0\,0)^T &: \left(0, -\frac{1}{\sqrt{3}}, \frac{1}{\sqrt{24}}, \frac{1}{\sqrt{40}}\right) \\
(0\,0\,0\,1\,0)^T &: \left(0, 0, -\sqrt{\frac{3}{8}}, \frac{1}{\sqrt{40}}\right) \\
(0\,0\,0\,0\,1)^T &: \left(0, 0, 0, -\sqrt{\frac{2}{5}}\right)
\end{aligned}
\tag{30}
$$

If all Abelian components of the chromomonopole condensate were of equal magnitude and mutually orthogonal in real space so that cross-terms could be neglected then they would all be confined at equal length scales and nothing remarkable would happen. However, we already know that such is not the case.

First consider the first three lines in equation (30). They all have identical (small) dependence on $\mathbf{H}^{(3)}, \mathbf{H}^{(4)}$, and the same dependencies on $\mathbf{H}^{(1)}, \mathbf{H}^{(2)}$ as in $SU(3)$ QCD. Consider now that the last two lines show no dependence on $\mathbf{H}^{(1)}, \mathbf{H}^{(2)}$, and it is only natural to equate the first three elements with the quark colours of the standard model. This is supported by noting that the two additional weight entries would provide very little additional confinement because the cross terms between $\mathbf{H}^{(3)}, \mathbf{H}^{(4)}$ have negative sign due to their antiparallelism. In fact the total contribution squared of $\mathbf{H}^{(3)}, \mathbf{H}^{(4)}$ is

$$
\left(\frac{1}{\sqrt{24}}\mathbf{H}^{(3)} + \frac{1}{\sqrt{40}}\mathbf{H}^{(4)}\right)^2 = H^2 \frac{1}{60}(4 - \sqrt{15}).
\tag{31}
$$

As can be seen from the bracket on the right-hand-side, the cross terms almost cancel this contribution entirely. According to the naïve interpretation of the dual superconductor model employed by this paper this corresponds to extremely weak confinement. Since it is inconsequential compared to the QCD confinement and might very well scale to zero at lower energies anyway (although I have not shown this!) I shall assume that this corresponds to an unconfined state.

This all ties in rather nicely with the result of section 4. in which the corresponding $SU(3)$ dynamics separate from the remaining dynamics through stronger scaling of the coupling constant.

The remaining weights have non-zero elements only in the third and fourth position. The final weight corresponds to a colour charge which I shall refer to as infrared ($\dot{r}$), whose confinement is slightly stronger than that of the QCD colours discussed above, while the penultimate one corresponds to the colour charge ultraviolet (w), whose confinement is nearly twice as strong as that of the QCD colours. (I shall refer to both of these charges as the invisible colours.) This occurs because there are positively contributing cross terms between $\mathbf{H}^{(3)}, \mathbf{H}^{(4)}$ (remember their antiparallelism). It follows that ultraviolet must be combined into some neutral combination with infrared at a very small length scale. Only combined with infrared does ultraviolet form an unconfined physical state.

Note that the third and fourth entries of the sum of the ultraviolet and infrared weights are exactly negative three times those of the QCD quark weights. We have already seen that such a combination effectively feels no confining effect from the background condensate. Remembering that the third and fourth Abelian directions provide the unconfined photon A_μ of section 5. gives the electric charge ratio between QCD quark states and white states. In conventional QCD the white states comprise both white combinations of QCD quark colours (hadrons) and truly colourless particles (leptons), and it is simply a fortunate coincidence that both have integer multiples of the electron charge. In this model, states carrying both of the invisible colours but no QCD colours are white and electrically charged, as are white combinations of the QCD colours. From the above discussion of the weights it follows that both these cases also have the same electric charge, up to a negative sign. It will therefore be natural to interpret the state with both invisible colours but no QCD colours as the electron/positron.

The reader may recall that the third and fourth Abelian directions also provide the confined photon E_μ. This may confuse some. While it may provide some additional interactions at close range, E_μ is *confined*, not *confining*, so the electric charges are still free to separate to very large distances.

Based on the proceeding discussion the fundamental representation can be shown as

$$[1] = (r\,b\,g\,\mathit{w}\,\mathring{r})^T, \tag{32}$$

where the electric charge of the QCD quarks is implicit in the colour. The red, blue and green quark colours are exchanged by the corresponding $SU(3)$ but the invisible colours are not exchanged except at extremely high energies due to the extra-strong confinement of the ultraviolet charge. At highest energies the dynamics are those of $SU(5)$ in the extreme weak-coupling limit. Electric charge has no meaning at such energies.

At intermediate energies at which the gluons XXXO have been confined out of the dynamics not only is there no available gluon to exchange the invisible colours, but the ultraviolet colour itself is confined just as QCD colours are confined at larger distances. This not only removes ultraviolet from the effective dynamics but the infrared as well, because the all-white combinations involving the infrared, apart from a meson-like bound state, require the ultraviolet. This confinement of infrared leaves no source for gluons of the form ??OX.

The intermediate dynamics consist primarily of conventional QCD and QED, but there is no obvious way to include the $SU(2)$ of weak nuclear decay or to turn a QCD quark into a lepton.

I now turn to the asymmetric representation

$$[2] = [1] \otimes_{AS} [1] = \begin{bmatrix} 0 & r/b & r/g & r/\mathit{w} & r/\mathring{r} \\ -r/b & 0 & b/g & b/\mathit{w} & b/\mathring{r} \\ -r/g & -b/g & 0 & g/\mathit{w} & g/\mathring{r} \\ -r/\mathit{w} & -b/\mathit{w} & -g/\mathit{w} & 0 & \mathit{w}/\mathring{r} \\ -r/\mathring{r} & -b/\mathring{r} & -g/\mathring{r} & -\mathit{w}/\mathring{r} & 0 \end{bmatrix}, \tag{33}$$

where [1] is the fundamental representation and $\otimes_{AS}$ indicates an antisymmetric cross-product. The top-left-hand corner has the same interpretation as in conventional GUT theories [28]. Red/blue is effectively antigreen etc, and the $U(1)$ (in this case electric) charge is double that of the QCD quarks in the fundamental representation. In other words the 3×3 block matrix in the top-left-hand corner can be associated with the anti-up quark when the fundamental representation contains the down quark. The remaining entries of [2] contain either an ultraviolet or an infrared colour charge, which confines them out of intermediate energy level dynamics. The one exception contains both invisible colour charges and is therefore a colourless state with electric charge negative three times that of the down quark, *ie.* a positron as discussed above. The effective dynamics of this representation, and its complex conjugate [3], are dominated by the colour and electric interactions of the up quark and electric interactions of the electron/positron.

7. Predictions and Prospects for Grand Unification

This approach yields no weak interaction dynamics but there is a gluon-mediated exchange between the up quarks and the electron. This is reminiscent of proton decay in conventional GUTs, in which a down quark becomes a positron and an up quark becomes an anti-up quark which forms a meson with the remaining up quark. In this case however, an up quark becomes an electron and the mediating gluon carries a charge of $+\frac{5}{3}$. This cannot be absorbed by anything within the proton so proton decay is forbidden. It could however be absorbed by an anti-up quark so that an extremely high energy collision between a proton and an anti-proton yielding an electron-positron pair and a neutral pair of pions, by which I mean either two π^0s or one π^+ and one π^-. Unfortunately such a sequence can also occur through ordinary standard model interactions so this is not much of a prediction.

It was natural to hope that dimensional frustration might yield a Higgsless GUT but it makes a good start, unifying $SU(3)_c$ with $U(1)_{EM}$, the forces acting on the right-handed matter fields, which do not feel the weak nuclear force. While I cannot yet claim to have done so, a physically realistic unification of $SU(3)_c$ with $U(1)_{EM}$ would be a unification of the forces affecting right-handed matter fields.

Assuming that nature does employ this mechanism to unify the strong nuclear and electromagnetic interactions of right-handed particles, what new phenomena could we expect to see? Obviously there would be new hadrons containing the invisible quarks. These can be divided into two types. There are mesons, we could call them invisible mesons, composed of invisible quark-antiquark pairs. Remember that such pairs can be taken not just from the fundamental representation which contains pure invisible states, but also from the [2] representation which contains states carrying both a QCD colour and an invisible colour. From QCD colour symmetry such states would be mixed, so the distinguishable invisible mesons are

$$w - \overline{w}, \; \dot{r} - \overline{\dot{r}}, \; w/\{rbg\} - \overline{w/\{rbg\}}, \; \dot{r}/\{rbg\} - \overline{\dot{r}/\{rbg\}}. \qquad (34)$$

It is, of course, possible that these states also mix, either with each other or with standard model mesons. This requires further study and is beyond the scope of this paper.

There is one more invisible meson which is listed separatedly because it is already known. Recall that the positron is the quark in the [2] representation with the colour charge

$uv/\dot{w}r$. The invisible meson associated with the positron is obviously positronium, so

$$e^+ - e^- \iff uv/\dot{w}r - \overline{uv/\dot{w}r}. \tag{35}$$

Another obvious combination is the quark pair made up of each invisible colour from the fundamental representation, $uv - \dot{w}r$, and of course its antimatter partner. Next there are particles made up of those quarks in [2] that carry both a QCD colour and an invisible colour, ie. $\{rbg\}/uv$ and $\{rbg\}/\dot{w}r$. An unconfined combination needs each QCD colour in equal numbers and each of the invisible colours in equal numbers, so at least six quarks are needed.

The quark combinations in the last paragraph are neither mesonic nor baryonic and may be candidates for dark matter. There may also be fancier combinations involving more quarks or even gluons similar to the exotic states discussed in relation to conventional QCD.

Conventional GUTs predict proton decay and monopole production. Dimensional frustration predicts neither of these. It does predict an anomolous scattering however. If an electron is fired into a proton at sufficiently high energies it may turn into an up quark and emit a mediating boson that is absorbed by an up quark and turns it into an electron. The experimenter sends in an electron and sees an electron emerge so this is just a scattering experiment, but it is additional scattering to the electromagnetic interaction already observed in deep inelastic scattering experiments. The strength of the scattering for a given electron energy has not yet been calculated and requires the mediating gluon mass which is currently unknown, although it is natural to expect that it is very massive so that extremely high energy scattering experiments would be needed. It is worth noting however that while reaching ever higher energies is becoming increasingly difficult, an experiment of this kind does not, by modern standards, require sophisticated detection equipment or calculations as it is only measuring electron scattering. Again, further work is required.

8. Conclusion

I have studied the long known but generally ignored result that QCD with five or more colours has an altered vacuum state due to the limited dimensionality of space, a condition dubbed 'dimensional frustration'. It appears to lead to a unified theory of the strong and electromagnetic interactions, which is not the conventional approach to grand unification, but these two forces are the only ones acting on the right-handed matter fields. Identification of the physical vacuum encounters an intractable set of non-analytic equations but further analysis was enabled by a well-motivated ansatz.

Assuming the dual superconductor model, a range of confinement scales emerged with one root vector (XXXO) being confined more strongly than all the rest, while some others are less tightly confined (??OO). Gluons remaining at intermediate energy scales exhibit unconventional dynamics because only some of them couple to the XXXO. An $SU(3)$ subset have stronger interactions among themselves at increasing length scales, suggesting the separation of QCD dynamics at lower energy. In addition to a weakly confined $U(1)$ and off-diagonal $SU(2)$ generators, there also emerged a single, unconfined $U(1)$ gauge field consistent with the photon. The theory appears to be a unification of QCD with electromagnetism. Such a unification, if consistent with experiment, is of interest to the standard model in which right-handed matter fields only couple to those two forces.

Study of the matter field representations found that the fundamental representation [1] comprises the three colours of down quark and two more so-called invisible colours, named ultraviolet and infrared. These two colour charges combined will neutralise each other, as do white combinations of the QCD colours. According to the area law of the Wilson loop [29] combined with the non-Abelian Stokes theorem [26, 27], the asymmetric arrangement of the chromomonopole condensate gives the invisible colours, especially ultraviolet, an extremely short confinement scale.

The antisymmetric combination of [1] with itself, denoted [2], contains the anti-up quark and the positron, as well as various combinations of the QCD and invisible colours. (The remaining matter representations [3], [4] are the complex conjugates of [1], [2].) Again the invisible-coloured quarks have a very short confinement scale and make little contribution to the intermediate energy dynamics so that the effective theory reduces to QCD and electromagnetism.

Much work remains to be done. The masses of gauge bosons coupling the QCD colours to the invisible colours need to be calculated. The breaking off of QCD symmetry, both through gluon interaction strength and quark confinement scales, suggests a strong symmetry breaking that should render these bosons very massive.

Dimensional frustration is a natural, almost inevitable, means of generating a hierarchy in QCD with five or more colours without resorting to contrived symmetry breaking methods such as the Higgs field. Even a simplistic analysis such as this finds a rich phenomenology.

The author thanks K.-I. Kondo for helpful discussions. This work was partially supported by a fellowship from the Japan Society for the Promotion of Science (P05717), with hospitality provided by the physics department of Chiba University.

References

[1] G.K. Savvidy. (1977) *Phys. Lett.*, **B71**:133.

[2] H. Flyvbjerg. (1980) *Nucl. Phys.*, **B176**:379.

[3] M.L. Walker. (2007) *JHEP*, **01**:056.

[4] Y. Nambu. (1974) *Phys. Rev.*, **D10**:4262.

[5] S. Mandelstam. (1976) *Phys. Rept.*, **23**:245-249.

[6] A.M. Polyakov. (1977) *Nucl. Phys.*, **B120**:429-458.

[7] Y.M. Cho. (1980) *Phys. Rev.*, **D21**:1080.

[8] G. 't Hooft. (1981) *Nucl. Phys.*, **B190**:455.

[9] L.D. Faddeev and A.J. Niemi. (1999) *Phys. Lett.*, **B449**:214-218.

[10] S.V. Shabanov. (1999) *Phys. Lett.*, **B463**:263-272.

[11] S. Li, Y. Zhang, and Z.-Y. Zhu. (2000) *Phys. Lett.*, **B487**:201-208.

[12] M.L. Walker. (2007) *arXiv:0707.2626 [hep-th]*, 2007.

[13] N.K. Nielsen and P. Olesen. (1978) *Nucl. Phys.*, **B144**:376.

[14] J. Honerkamp. (1972) *Nucl. Phys.*, **B48**:269-287.

[15] Y.M. Cho, M.L. Walker, and D.G. Pak. (2004) *JHEP*, **05**:073.

[16] Y.M. Cho and M.L. Walker. (2004) *Mod. Phys. Lett.*, **A19**:2707-2716.

[17] Y.M. Cho and D.G. Pak. (2002) *Phys. Rev.*, **D65**:074027.

[18] K.-I. Kondo. (2004) *Phys. Lett.*, **B600**:287-296.

[19] D. Kay, A. Kumar, and R. Parthasarathy. (2005) *Mod. Phys. Lett.*, **A20**:1655-1662.

[20] P.H. Frampton. *Quantum Field Theories*. (1987) Benjamin-Cummings, California.

[21] S. Kato et al. (2006) *Phys. Lett.*, **B632**:326-332.

[22] Y. Hatta and K. Fukushima. (2004) *Phys. Rev.*, **D69**:097502.

[23] F. Karsch, E. Laermann, and A. Peikert. (2001) *Nucl. Phys.*, **B605**:579-599.

[24] L.G. Yaffe and B. Svetitsky. (1982) *Phys. Rev. D*, **26**(4):963-965.

[25] R.D. Pisarski and F. Wilczek. (1984) *Phys. Rev.*, **D29**:338-341.

[26] K.-I. Kondo and Y. Taira. (2000) *Mod. Phys. Lett.*, **A15**:367-377.

[27] K.-I. Kondo and Y. Taira. (2000) *Prog. Theor. Phys.*, **104**:1189-1265.

[28] H. Georgi. *Lie Algebras in Particle Physics:From Isospin to Unified Theories*. (1999) Westview Press, Boulder, Colorado.

[29] K.G. Wilson. (1974) *Phys. Rev.*, **D10**:2445-2459.

In: High Energy Physics Research Advances
Editors: T.P. Harrison et al, pp. 95-110

Chapter 4

AdS/CFT Approach to Melvin Field Deformed Wilson Loop

Wung-Hong Huang[*]
Department of Physics, National Cheng Kung University
Tainan, Taiwan

Abstract

We first apply the transformation of mixing azimuthal and internal coordinate to the 11D M-theory with a stack N M2-branes to find the spacetime of a stack of N D2-branes with Melvin one-form in 10D IIA string theory, after the Kaluza-Klein reduction. Next, we apply the Melvin twist to the spacetime and perform the T duality to obtain the background of a stack of N D3-branes. In the near-horizon limit the background becomes the Melvin field deformed $AdS_5 \times S^5$ with NS-NS B field. In the AdS/CFT correspondence it describes a non-commutative gauge theory with non-constant non-commutativity in the Melvin field deformed spacetime. We study the Wilson loop therein by investigating the classical Nambu-Goto action of the corresponding string configuration and find that, contrast to that in the undeformed spacetime, the string could be localized near the boundary. Our result shows that while the geometry could only modify the Coulomb type potential in IR there presents a minimum distance between the quarks. We argue that the mechanism behind producing a minimum distance is coming from geometry of the Melvin field deformed background and is not coming from the space non-commutativity.

1. Introduction

The expectation value of Wilson loop is one of the most important observations in the gauge theory. In the AdS/CFT duality [1-3] it becomes tractable to understand this highly nontrivial quantum field theory effect through a classical description of the string configuration in the AdS background. Using this AdS/CFT duality Maldacena [4] derived for the first time the expectation value of the rectangular Wilson loop operator from the Nambu-Goto action associated with the $AdS_5 \times S^5$ supergravity and found that the interquark potential exhibits the Coulomb type behavior expected from conformal invariance of the gauge theory.

---
[*]E-mail address: whhwung@mail.ncku.edu.tw

Maldacena's computational technique has already been extended to the finite temperature case by replacing the AdS metric by a Schwarzschild-AdS metric [5]. In order to make contact with Nature many investigations had gone beyond the initial conjectured duality and generalized the method to investigate the theories breaking conformality and (partially) supersymmetry [6-9]. For example, the Klebanov–Witten solution [10], the Klebanov–Tseytlin solution [11], the Klebanov–Strassler solution [12], and Maldacena–Núñez (MN) solution [13] which dual to the $\mathcal{N} = 1$ gauge theory.

The technique has also been used to investigate the Wilson loop in non-commutative gauge theories. First, the dual supergravity description of non-commutative gauge theory with a constant non-commutativity was constructed by Hashimoto and Itzhaki [14]. The corresponding string background is the curved spacetime where the constant B field is turned on along the brane worldvolume. The associated Wilson loop studied by Maldacena and Russo [15] showed that the string cannot be located near the boundary and we need to take a non-static configuration in which the quark-antiquark acquire a velocity on the non-commutative space. The result shows that the interquark potential exhibits the Coulomb type behavior as that in commutative space.

In [16] we use AdS/CFT correspondence to investigate the Wilson loop on the non-commutative gauge theory with a non-constant non-commutativity. The dual supergravity is the Melvin-Twist deformed $AdS_5 \times S^5$ background which was first constructed by Hashimoto and Thomas [17]. The corresponding string is on the curved background where the non-constant B field is turned on along the brane worldvolume. After analyzing the Nambu-Goto action of the classical string configuration we had shown that [16], while the non-commutativity could modify the Coulomb type potential in IR it may produce a strong repulsive force between the quark and anti-quark if they are close enough. In particular, we show that there presents a minimum distance between the quarks, which is proportional to the value of the non-commutativity. In this paper we will extend the previous paper to study the non-commutative gauge theory in a Melvin field deformed spacetime in the dual string description.

In section II we first apply the transformation of mixing azimuthal and internal coordinate [18] to the 11D M-theory with a stack N M2-branes [19] to find the spacetime of a stack of N D2-branes with Melvin one-form in 10D IIA string theory, after the Kaluza-Klein reduction. Next, we apply the Melvin twist to the spacetime and perform the T duality [20] to obtain the background of a stack of N D3-branes. In the near-horizon limit the background becomes the Melvin field deformed $AdS_5 \times S^5$ with NS-NS field. In the AdS/CFT correspondence it describes the non-commutative gauge theory in the Melvin field deformed spacetime.

In section III we investigate the Wilson loop therein by dual string description. We find that, contrast to that in the undeformed space, the string could be localized near the boundary. Our result shows that there presents a minimum distance between the quarks. We argue that the mechanism of producing a minimum distance is coming from geometry of the Melvin field deformed background, contrast that in [16] which is coming from the space non-commutativity. In section IV we discuss the particle trajectory on the Melvin field deformed $AdS_5 \times S^5$ background and see that the property of the Melvin field effect on the Wilson loop is, more or less, similar to that on the particle trajectory. The last section is devoted to a short discussion.

2. Supergravity Dual of Non-commutative Gauge Theory on Melvin Field Deformed Spacetime

We first apply the transformation of mixing azimuthal and internal coordinate [18] to the 11D M-theory with a stack N M2-branes [19] and then apply the Kaluza-Klein reduction to find the spacetime of a stack of N D2-branes with one-form in 10D IIA string theory.

The full N M2-branes solution is given by

$$ds^2_{11} = H^{\frac{-2}{3}}\left(-dt^2 + dr^2 + r^2 d\phi^2\right) + H^{\frac{1}{3}}\left(dz^2 + dU^2 + U^2 d\Omega_5^2 + dx_{11}^2\right), \quad (2.1)$$

$$A^{(3)} = H^{-1}dt \wedge dx_1 \wedge dx_2. \quad (2.2)$$

H is the harmonic function defined by

$$H = 1 + \frac{R}{r^{D-p-3}}, \quad r^2 \equiv z^2 + \rho^2 + x_{11}^2, \quad R \equiv \frac{16\pi G_D T_p N}{D-p-3}, \quad (2.3)$$

in which G_D is the D-dimensional Newton's constant and T_p the p-brane tension. In the case of (2.1), $D = 11$ and $p = 2$.

We first transform the angle ϕ by mixing it with the compactified coordinate x_{11} in the following substituting

$$\phi \rightarrow \phi + Cx_{11}. \quad (2.4)$$

Using the above substitution the line element (2.1) becomes

$$ds^2_{11} = H^{\frac{-2}{3}}\left(-dt^2 + dr^2\right) + H^{\frac{1}{3}}\left(dz^2 + dU^2 + U^2 d\Omega_5^2\right) + \left(r^2 H^{\frac{-2}{3}} - \frac{C^2 r^4 H^{\frac{-4}{3}}}{H^{\frac{1}{3}} + C^2 r^2 H^{\frac{-2}{3}}}\right)d\phi^2$$

$$+ \left(H^{\frac{1}{3}} + C^2 H^{\frac{-2}{3}}\right)\left[dx_{11} + \frac{Cr^2 H^{\frac{-2}{3}}d\phi}{H^{\frac{1}{3}} + C^2 r^2 H^{\frac{-2}{3}}}\right]^2. \quad (2.5)$$

Using the relation between the 11D M-theory metric and string frame metric, dilaton field and Melvin 1-form potential

$$ds^2_{11} = e^{-2\phi/3}ds^2_{10} + e^{4\phi/3}(dx_{11} + 2A_\mu dx^\mu)^2, \quad (2.6)$$

the 10D IIA background is described by

$$ds^2_{10} = \sqrt{1 + C^2 r^2 H^{-1}}\left[H^{\frac{-1}{2}}\left(-dt^2 + dr^2 + \frac{r^2 d\phi^2}{1 + C^2 r^2 H^{-1}}\right) + H^{\frac{1}{2}}\left(dz^2 + dU^2 + U^2 d\Omega_5^2\right)\right]. \quad (2.7)$$

$$A_\phi = \frac{Cr^2}{2\left(H + C^2 r^2\right)}, \quad (2.8)$$

in which A_ϕ is the RR one-form potential. The dilaton field and other RR potential arisen from RR there-form $A^{(3)}$ will be neglected hereafter as they are irrelevant to our calculation of the Wilson loop in section III. In the case of $C = 0$ the above spacetime becomes the well-known geometry of a stack of N D2-branes. Thus, the background describes the spacetime of a stack of N D2-branes with Melvin field flux.

We next transform the angle ϕ by mixing it with the coordinate z in the following substituting

$$\phi \rightarrow \phi + Cz. \tag{2.9}$$

Using the above substitution the line element (2.7) becomes

$$ds_{10}^2 = \sqrt{1 + C^2 r^2 H^{-1}} \left[H^{\frac{-1}{2}} \left(-dt^2 + dr^2 + \frac{r^2 d\phi^2}{1 + C^2 r^2 H^{-1}} \right) + \frac{2Br^2 H^{\frac{-1}{2}} d\phi dz}{1 + C^2 r^2 H^{-1}} \right.$$

$$\left. + \left(\frac{B^2 r^2 H^{\frac{-1}{2}}}{1 + C^2 r^2 H^{-1}} + H^{\frac{1}{2}} \right) dz^2 + H^{\frac{1}{2}} \left(dU^2 + U^2 d\Omega_5^2 \right) \right]. \tag{2.10}$$

Finally, we perform the T-duality transformation [20] on the coordinate z. The result supergravity background becomes

$$ds_{10}^2 = \sqrt{1 + C^2 r^2 H^{-1}} \left[H^{\frac{-1}{2}} \left(-dt^2 + dr^2 + \frac{r^2 d\phi^2 + dz^2}{1 + C^2 r^2 H^{-1} + B^2 r^2 H^{-1}} \right) + H^{\frac{1}{2}} (dU^2 + U^2 d\Omega_5^2) \right]. \tag{2.11}$$

$$A_{\phi z} = \frac{Cr^2}{2(H + C^2 r^2)}, \qquad B_{\phi z} = \frac{Br^2 H^{-1}}{1 + C^2 r^2 H^{-1} + B^2 r^2 H^{-1}}, \tag{2.12}$$

in which $A_{\phi z}$ is the RR two-form potential and $B_{\phi z}$ the NS-NS B field.

Using above two equations we have two results:
(1) In the near-horizon limit Eqs. (2.11) and (2.12) become

$$ds_{10}^2 = \sqrt{1 + C^2 r^2 U^4} \left[U^2 \left(-dt^2 + dr^2 + \frac{r^2 d\phi^2 + dz^2}{1 + C^2 r^2 U^4 + B^2 r^2 U^4} \right) + U^{-2} (dU^2 + U^2 d\Omega_5^2) \right]. \tag{2.13}$$

$$A_{\phi z} = \frac{Cr^2 U^4}{2(1 + C^2 r^2 U^4)}, \qquad B_{\phi z} = \frac{Br^2 U^4}{1 + C^2 r^2 U^4 + B^2 r^2 U^4}. \tag{2.14}$$

This is the Melvin field deformed $AdS_5 \times S^5$ with NS-NS B field. The Melvin field here means the RR two-form potential $A_{\phi z}$ which arises from the Melvin twist of (2.4). The NS-NS B field is $B_{\phi z}$ which arises from the Melvin twist of (2.9).
(2) Taking $H = 1$ then Eqs. (2.11) and (2.12) become

$$ds_{10}^2 = \sqrt{1 + C^2 r^2} \left[\left(-dt^2 + dr^2 + \frac{r^2 d\phi^2 + dz^2}{1 + C^2 r^2 + B^2 r^2} \right) + (dU^2 + U^2 d\Omega_5^2) \right]. \tag{2.15a}$$

$$B_{\phi z} = \frac{Br^2}{1 + C^2 r^2 + B^2 r^2}. \tag{2.15b}$$

Using the above IIB background we can find a gauge theory by applying the mapping of Seiberg and Witten [21]

$$(G + \theta)^{\mu\nu} = [(g + B)_{\mu\nu}]^{-1}. \tag{2.16}$$

The dual 4D gauge theory is thus on the Melvin field deformed spacetime with the line element

$$G_{\mu\nu}dx^\mu dx\nu = \sqrt{1 + C^2 r^2} \left[-dt^2 + dr^2 + \frac{r^2 d\phi^2 + dz^2}{1 + C^2 r^2} \right], \qquad (2.17a)$$

with space non-commutativity

$$\theta^{\phi z} = B. \qquad (2.17b)$$

When $C = 0$ above result is just that described by Hashimoto and Thomas [17]. Note that replacing the coordinate (t, r, ϕ, z) by Cartesian coordinates (t, x, y, z) the noncommutativity become $\theta^{xz} = -yB$ and $\theta^{yz} = xB$ [17]. Our background thus describes the non-commutative gauge theory on the Melvin field deformed spacetime with a non-constant non-commutativity.

3.   Wilson Loop of Non-commutative Gauge theory on Melvin Field Deformed Spacetime

3.1.   Formulation

Following the Maldacena's computational technique the Wilson loop of a quark anti-quark pair is calculated from a dual string. The string lies along a geodesic with endpoints on the AdS_5 boundary representing the quark and anti-quark positions. The ansatz for the background string we will consider is

$$t = \tau, \quad z = \sigma, \quad U = U(\sigma), \qquad (3.1)$$

and rest of the string position is constant in σ and τ. We choose $r = r_0 = 1$ for a convenience. The Nambu-Goto action becomes

$$S = \frac{T}{2\pi} \int d\sigma \sqrt{(1 + C^2 U^4)(\partial_\sigma U)^2 + \frac{U^4(1 + C^2 U^4)}{1 + C^2 U^4 + B^2 U^4}}, \qquad (3.2)$$

in which T denotes the time interval we are considering and we have set $\alpha' = 1$. As the associated Lagrangian $(\mathcal{L})$ does not explicitly depend on σ the relation $(\partial_\sigma U)\frac{\partial \mathcal{L}}{\partial(\partial_\sigma U)} - \mathcal{L}$ will be proportional to an integration constant U_0. This implies the following relation

$$\frac{\frac{U^4(1+C^2 U^4)}{1+C^2 U^4+B^2 U^4}}{\sqrt{(1 + C^2 U^4)(\partial_\sigma U)^2 + \frac{U^4(1+C^2 U^4)}{1+C^2 U^4+B^2 U^4}}} = constant. \qquad (3.3)$$

To describe a quark pair with a finite distance the dual string configuration shall satisfy the following boundary condition

$$\partial_\sigma U \to \infty, \quad as \quad U \to \infty. \qquad (3.4)$$

In the case of $C = 0$ above condition implies that the *constant* in (3.3) is zero and we have only null solution of $U = 0$. This is what Maldacena and Russo [15] had showed that the string cannot be located near the boundary and we need to take a non-static configuration in which the quark-antiquark acquires a velocity on the non-commutative space. The case of $C = 0$ had been analyzed by us in a previous paper [16], in which we considered the case of $\phi = vt$. The dual string thus is moving with a constant angular velocity v.

In the case of $C \neq 0$ the *constant* in (3.3) could be a finite value which may be calculated from another condition

$$\partial_\sigma U = 0, \quad as \quad U = U_0, \tag{3.5}$$

in which U_0 is the minimum of U of the dual string configuration. In this case (3.3) implies

$$\frac{dU}{d\sigma} = \pm \frac{U^2}{U_0^2} \frac{1}{\sqrt{1+(C^2+B^2)U^4}} \sqrt{U^4 \frac{1+C^2U^4}{1+(C^2+B^2)U^4} \frac{1+(C^2+B^2)U_0^4}{1+C^2U_0^4} - U_0^4}. \tag{3.6}$$

As we put the quark at place $z = \sigma = -L/2$ and the anti-quark at $z = \sigma = L/2$ we can use the above relation to find the following relation between the integration constant U_0 and L.

$$L/2 = \int_0^{L/2} d\sigma = \int_{U_0}^\infty dU(\partial_\sigma U)^{-1} = \frac{1}{U_0} \int_0^1 \frac{dx}{x^{\frac{3}{4}}} \frac{\sqrt{x+(C^2+B^2)U_0^4}}{\sqrt{\frac{x+C^2U_0^4}{x+(C^2+B^2)U_0^4} \frac{1+(C^2+B^2)U_0^4}{1+C^2U_0^4} - x}}. \tag{3.7}$$

In the same way, using (3.6) the interquark potential evaluated from the action (3.2) becomes

$$H = \frac{U_0}{4\pi} \left(\int_1^\infty dy \left[\frac{\sqrt{y^4 \frac{1+C^2U_0^4y^4}{1+(C^2+B^2)U_0^4y^4} \frac{1+(C^2+B^2)U_0^4}{1+C^2U_0^4}}}{\sqrt{y^4 \frac{1+C^2U_0^4y^4}{1+(C^2+B^2)U_0^4y^4} \frac{1+(C^2+B^2)U_0^4}{1+C^2U_0^4} - 1}} y^\epsilon - y^\epsilon \right] - 1 \right)$$

$$= \frac{U_0}{4\pi} \int_0^1 dx x^{-(\frac{7}{4}+\frac{\epsilon}{4})} \sqrt{x + C^2U_0^4} \frac{\sqrt{\frac{x+C^2U_0^4}{x+(C^2+B^2)U_0^4} \frac{1+(C^2+B^2)U_0^4}{1+C^2U_0^4}}}{\sqrt{\frac{x+C^2U_0^4}{x+(C^2+B^2)U_0^4} \frac{1+(C^2+B^2)U_0^4}{1+C^2U_0^4} - x}}, \tag{3.8}$$

in which we have followed the prescription of Maldacena [4] by multiplying the integration a factor y^ϵ and subtraction the regularized mass of W-boson to find the finite result. We can use the Eqs.(3.7) and (3.8) to analyze the string configuration on the Melvin field deformed background with NS-NS B field. The result could be used to find the interquark potential for the system on the deformed background with a non-commutativity.

3.2. Analyses: Interquark Distance

For a clear illustration we show in figure 1 the function $L(U_0)$ which is found by performing the numerical evaluation of (3.7) for the cases of $B = 0.1$ with $C = 0.01$ and $C = 0.1$ respectively. Figure 2 shows the function $L(U_0)$ for the cases of $C = 0.01$ with $B = 0.1$ and $B = 0.2$ respectively. The dashed line therein represents that with $C = B = 0$.

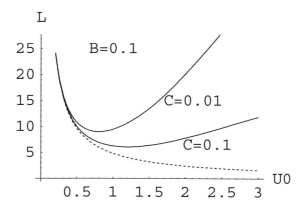

Figure 1. The function $L(U_0)$ for the cases of $B = 0.1$ with $C = 0.01$ and $C = 0.1$ respectively. The dashed line represents that with $C = B = 0$.

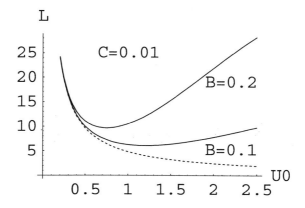

Figure 2. The function $L(U_0)$ for the cases of $C = 0.01$ with $B = 0.1$ and $B = 0.2$ respectively. The dashed line represents that with $C = B = 0$.

Figure 1 shows that there presents a minimum distance between the quarks and that the minimum distance is a decreasing function of C.

Figure 2 shows that there presents a minimum distance between the quarks and that the minimum distance is an increasing function of B.

These properties could be read from the following analyses. Using (3.7) we have an approximation

$$L \approx \frac{4(2\pi)^{3/2}}{\Gamma(1/4)^2}\frac{1}{U_0} + \frac{2B^2\Gamma(1/4)^2}{\sqrt{\pi}\sqrt{C}}U_0^2, \quad if \quad C \leq B \neq 0, \tag{3.9}$$

which implies a minimum distance

$$L_0 \approx \frac{2^{2/3}4B^{2/3}\pi^{1/6}}{C^{1/2}}\left(\frac{4\pi}{\Gamma(1/4)}\right)^{2/3}, \quad if \quad C \leq B \neq 0. \tag{3.10}$$

Above result is the approximation of small values of B and C and shows that while the minimum value of the interquark distance L_0 is a decreasing function of C it is an increasing function of B. Note that Eq.(3.10) implies that $L_0 \to \infty$ as $C \to 0$. This is because that when $C \to 0$ the string could not be localized near the boundary and we need to consider a moving string configuration as that analyzed in our previous paper [16]. Thus the above result is not an analytic function at $C = 0$.

Although the previous paper [16] had also found that the space noncommutativity will lead the interquark distance L to be larger than L_0 the mechanism therein and that in this paper does not have a similar origin. In [16] we consider the case of $C = 0$ and find that, contrast to the figure 1, the distance L is a decreasing function of U. It then asymptotically approach to the minimum distance L_0. The property of asymptotically approaching to the minimum distance L_0 means that it will exist an extremely repulsive when the quark distance is approaching to L_0. *The repulsive force is coming from the space non-commutativity.* Figures 1 and 2, however, show that distance L could become L_0 at finite value of U_0, which means that mechanism of producing a minimum distance in this paper is not coming from the space non-commutativity. In fact it arises from the Melvin field which render the theory to be in a deformed background and *geometry of the background plays a role to produce this minimum distance L_0.*

To explicitly see this property we may analyze the system of $B = 0$. In this case we can plot the function $L(U_0)$ by performing the numerical evaluation of (3.7). The result is shown in figure 3 for the cases of $C = 0, 1$ and 2 respectively.

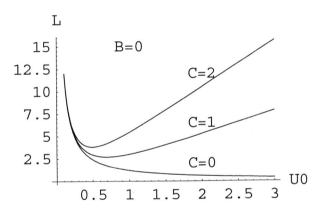

Figure 3. The function $L(U_0)$ for the cases of $B = 0$ with $C = 0, 1$ and 2 respectively.

It then see that the distance L becomes L_0 at finite value of U_0, as that in figure 1. However, contrast to the figure 1, the minimum distance is an increasing function of C. These properties could be read from the following analyses. In the case of $B = 0$ Eq.(3.7) gives

$$L \approx \frac{4(2\pi)^{3/2}}{\Gamma(1/4)^2} \frac{1}{U_0} + \frac{4\Gamma(1/4)^2}{3\sqrt{\pi}} C^{3/2} U_0^2, \tag{3.11}$$

which implies a minimum distance

$$L_0 \approx \frac{8\,\Gamma(1/4)^2}{3\sqrt{\pi}} C^{1/2}. \tag{3.12}$$

The minimum distance is an increasing function of C. Thus we conclude that the mechanism behind producing a minimum distance is coming from geometry of the Melvin field deformed background and is not coming from the space non-commutativity.

It is worthy to mention that, although (3.10) shows that the minimum distance is a decreasing function of C it is only an good approximate for $B \geq C$. In fact, when C is larger than B the minimum distance will be an increasing function of C. The behavior is like that in figure 3 and could be seen from the following analyses. In the case of $C \gg B$ Eq,(3.7) gives an approximation

$$L \approx \frac{4(2\pi)^{3/2}}{\Gamma(1/4)^2} \frac{1}{U_0} + \frac{4\Gamma(1/4)^2}{3\sqrt{\pi}} C^{3/2} U_0^2 + \frac{4(2\pi)^{3/2}}{\Gamma(1/4)^2} B^2 U_0^3, \quad if \ \ C \gg B, \quad (3.13)$$

which implies a minimum distance

$$L_0 \approx \frac{8\,\Gamma(1/4)^2}{3\sqrt{\pi}} C^{1/2} + \frac{3}{\sqrt{2}} \left(\frac{4(2\pi)^{3/2}}{\Gamma(1/4)^2} \right)^{4/3} \frac{B^2}{C}, \quad if \ \ C \gg B. \quad (3.14)$$

Thus the minimum distance is an increasing function of C. Note that when $B = 0$ then (3.13) and (3.14) reduce to (3.11) and (3.12) respectively.

Finally, from figures 1, 2 and 3 we see that above L_0 there have two values of U_0 which could correspond to a distance L. In fact, the solution of large value U_0 will have larger energy than that of small value U_0, as analyzed in bellow.

3.3.   Analyses: Interquark Potential

Using (3.8) we have approximations

$$H \approx -\frac{\sqrt{2\pi}}{\Gamma(1/4)^2} \left[\frac{4(2\pi)^{3/2}}{\Gamma(1/4)^2} \frac{1}{L} + \frac{2\Gamma(1/4)^2}{\sqrt{\pi}} \left(\frac{4(2\pi)^{3/2}}{\Gamma(1/4)^2} \right)^3 \frac{B^2}{\sqrt{C}} \frac{1}{L^4} \right], \quad if \ \ C \leq B.$$

$$(3.15)$$

$$H \approx -\frac{\sqrt{2\pi}}{\Gamma(1/4)^2} \left[\frac{4(2\pi)^{3/2}}{\Gamma(1/4)^2} \frac{1}{L} + \frac{4\Gamma(1/4)^2}{3\sqrt{\pi}} \frac{C^{3/2}}{L^3} + \left(\frac{4(2\pi)^{3/2}}{\Gamma(1/4)^2} \right)^5 \frac{B^2}{\sqrt{L^5}} \right], \quad if \ \ C \gg B.$$

$$(3.16)$$

To obtain the final result we have used the relations (3.9) and (3.13) respectively. The Melvin field will therefore slightly modify the Coulomb potential in IR.

In the case of large U_0 Eqs.(3.7) and (3.8) imply the relations

$$L \approx \frac{2\Gamma(1/4)^2}{\sqrt{2\pi}} \sqrt{C^2 + B^2} U_0, \quad (3.17)$$

$$H \approx \frac{\Gamma(1/4)^2}{12\pi\sqrt{2\pi}} C U_0^3 \approx \frac{1}{48\Gamma(1/4)^4} \frac{C}{(C^2 + B^2)^{3/2}} L^3. \quad (3.18)$$

Thus the the solution of large value U_0 will have larger energy than that of small value U_0. Finally, we show in figure 4 the interquark potential $H(L)$ of the non-commutative gauge theory on the Melvin-field deformed spacetime with non-constant non-commutativity.

According to the AdS/CFT duality the property of gauge theory is calculated from the dual string on the AdS space. The geometric property of the AdS shall be relevant to the

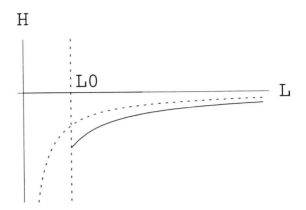

Figure 4. The interquark potential $H(L)$. We see that there presents a minimum distance L_0 between the quarks and it becomes Coulomb phase potential in IR. The dashed line represents that with $C = B = 0$.

character the gauge theory. As the string corresponding to the Wilson loop lies along a geodesic with endpoints on the AdS_5 boundary the geometry of the AdS_5 therefore will bend the string to be a "U" type. For the gauge theory under the Melvin field deformation the corresponding geometry is the Melvin field deformed AdS spacetime and the geometry of the deformed AdS_5 will bend the string to be a "U" type, which in our model has shown a property of existing a minimum distance L_0 between the quarks. Therefore, it is interesting to see whether such a property will also be shown in the trajectory of a particle. In the section IV we will investigate this problem and see that the effect of the Melvin field on the particle is, more or less, similar to that on the string.

4.   Particle Trajectory on Melvin Field Deformed AdS

In this section we will analyze the motion of a particle under the Melvin field deformed spacetime described in (2.13). As discussed in section III the minimum distance L_0 in our model is produced by the geometry of the background and not from the non-commutativity, therefore we will analyze the particle trajectory on the background with $B = 0$ and see what difference between that with $C = 0$ and that with $C \neq 0$.

We will consider the following metric

$$ds_{10}^2 = \sqrt{1 + C^2 U^4} \left[U^2 \left(-dt^2 + \frac{dx^2}{1 + C^2 U^4} \right) + U^{-2} dU^2 \right]. \tag{4.1}$$

in which x is used to denote the coordinate z in (2.13).

4.1.   Particle Trajectory on Undeformed AdS Background

Let us first analyze the undeformed spacetime with $C = 0$. The equations of motion, $0 = \ddot{X}^\mu + \Gamma^\mu_{\alpha\beta} \dot{X}^\alpha \dot{X}^\beta$, lead to the three equations

$$0 = \ddot{t} + \frac{2}{U} \dot{t} \dot{U}, \tag{4.2}$$

$$0 = \ddot{x} + \frac{2}{U} \, \dot{x} \, \dot{U}, \tag{4.3}$$

$$0 = \ddot{U} + U^3 \dot{t}^2 - U^3 \dot{x}^2 - U^{-1} \dot{U}^2, \tag{4.4}$$

in which $\dot{t} \equiv dt/d\tau$ and τ is the proper time, and so on. In the case of $B = 0$ Eq.(4.1) implies

$$1 = -U^2 \dot{t}^2 + U^2 \, \dot{x}^2 + U^{-2} \dot{U}^2. \tag{4.5}$$

Equations (4.2) and (4.3) give the solutions

$$\dot{t} = (\dot{t})_0 \left(\frac{U_0}{U} \right)^2, \tag{4.6}$$

$$\dot{x} = (\dot{x})_0 \left(\frac{U_0}{U} \right)^2, \tag{4.7}$$

in which $U_0 \equiv U(\tau = 0)$, and so on. Substituting (4.6) and (4.7) into (4.5) we have a relation

$$\dot{U} = \sqrt{(\dot{t})_0^2 \, U_0^2 - (\dot{x})_0^2 \, U_0^2 + U^2}. \tag{4.8}$$

From (4.7) and (4.8) we can find an useful differential equation

$$\frac{dx}{dU} = \frac{(\dot{x})_0 U_0^2}{\sqrt{(\dot{t})_0^2 \, U_0^2 - (\dot{x})_0^2 \, U_0^2 + U^2}} \frac{1}{U^2}, \tag{4.9}$$

which has a simple solution

$$x(U) - x_0 = \frac{(\dot{x})_0}{(\dot{t})_0^2 - (\dot{x})_0^2} \left(\sqrt{(\dot{t})_0^2 - (\dot{x})_0^2 + 1} - \frac{\sqrt{(\dot{t})_0^2 \, U_0^2 - (\dot{x})_0^2 \, U_0^2 + U^2}}{U} \right). \tag{4.10}$$

Above solution tells us that the particle trajectory on the AdS spacetime is "bended". For a clear illustration we show in figure 5 the particle trajectory on AdS.

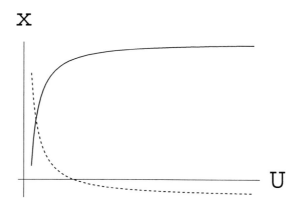

Figure 5. The particle trajectory on AdS. Note that $x(U \to \infty) \to a \ finite \ constant$.

It shall be noticed that we do not want to show the particle trajectory will be bent into a "U" type. What we attempt to see is the property that the "bending" property shown in the particle trajectory is coming from the geometric character of the background AdS spacetime which is that to bend the string to lies along a geodesic with a "U" type. In fact, from the figure 3 we see that a particle trajectory along a dashing line will approach to $x \to \pm\infty$. Thus there is a spiky on the cross point of the dashing line and solid line. However, the string geodesic is a smooth "U" type without any spiky.

4.2. Particle Trajectory on Deformed AdS background

We next analyze the deformed spacetime with $C \neq 0$. In the case of $C \gg 1$ The equations of motion become

$$0 \approx \ddot{t} + \frac{4}{U} \dot{t}\dot{U},$$
(4.11)

$$0 \approx \ddot{x},$$
(4.12)

$$0 \approx \ddot{U} + 2U^3\dot{t}^2.$$
(4.13)

In this case (4.1) implies the relation

$$1 \approx -CU^4\dot{t}^2 + \frac{1}{C}\dot{x}^2 + C\dot{U}^2.$$
(4.14)

Equations (4.11) and (4.12) give the solutions

$$\dot{t} = (\dot{t})_0 \left(\frac{U_0}{U}\right)^4.$$
(4.15)

$$\dot{x} = (\dot{x})_0.$$
(4.16)

Substituting (4.15) and (4.16) into (4.14) we have a relation

$$\dot{U} \approx \sqrt{\frac{1}{C}\left[1 + \frac{C(\dot{t})_0^2 U_0^8}{U^4}\right]}.$$
(4.17)

From (4.16) and (4.17) we can find a useful differential equation

$$\frac{dx}{dU} = \frac{(\dot{x})_0\sqrt{C}}{\sqrt{1 + C(\dot{t})_0^2 U_0^8 U^{-4}}},$$
(4.18)

which implies the following relation

$$x(U) - x_0 = \sqrt{C}\int_{U_0}^{U} \frac{(\dot{x})_0\, dU}{\sqrt{1 + C(\dot{t})_0^2 U_0^8 U^{-4}}}.$$
(4.19)

Thus

$$x(U \to \infty) \to \infty,$$
(4.20)

contrast to that in the undeformed system in which $x(U \to \infty) \to a\ finite\ constant$. This means that the geometry of the Melvin field deformed space has an incline to increasing the distance of final position $x(U \to \infty)$ from the initial position x_0. Note that this property comes from the geometry of the Melvin field deformed background and one may conjecture that it will also be shown in the string configuration which corresponding to the Wilson loop lies along a geodesic with endpoints on the AdS_5 boundary. We may argue that the corresponding property is that shown in (3.12), i.e. interquark minimum distance L_0 is an increasing function of $\sqrt{C}$.

5. Conclusion

In this paper we study the Wilson loop in the Melvin field deformed $AdS_5 \times S^5$ background with NS-NS B field from the supergravity side of the duality proposed by Maldacena [4]. We first construct the background through a series of Melvin twist and a T-dual transformation on the 11D M-theory after the Kaluza-Klein reduction. In the AdS/CFT correspondence it describes the non-commutative gauge theory in the Melvin field deformed spacetime. We investigate the Wilson loop therein by dual string description. We have found that, contrast to that in the undeformed space, which was investigate by us in a previous paper [16], the string could be localized near the boundary. Our result show that there presents a minimum distance between the quarks. We argue that the mechanism of producing a minimum distance is not coming from the space non-commutativity but form the geometry of the Melvin field deformed background. We also have discussed the particle trajectory on the Melvin field deformed $AdS_5 \times S^5$ background and see that the property of the Melvin field effect on the Wilson loop is, more or less, similar to that on the particle trajectory.

Finally, it is known that, Maldacena method [4] cannot obtain the subleading corrections which arise when one considers coincident Wilson loops, multiply wound Wilson loops or Wilson loops in a higher dimensional representation. In recent, Drukker and Fiol [22] showed a possible way to compute a class of these loops using D branes carrying a large fundamental string charge dissolved on their worldvolume pinching off at the boundary of the AdS on the Wilson loop [23]. It is interesting to use the D-brane approach to evaluate the Melvin field deformed Wilson loop of other class. Another interesting phenomena in the AdS picture of Wilson loops is the Gross-Ooguri phase transition [24] which occurs in the two Wilson loop correlator [25-27]. It would be interesting to see how the Melvin field would affect the Gross-Ooguri phase transition therein. These problems remain in the future investigations.

References

[1] J. M. Maldacena, "The large N limit of superconformal field theories and supergravity," *Adv. Theor. Math. Phys.* **2** (1998) 231-252 [hep-th/9711200].

[2] E. Witten, "Anti-de Sitter space and holography," *Adv. Theor. Math. Phys.* **2** (1998) 253 [hep-th/9802150].

[3] S. S. Gubser, I. R. Klebanov and A. M. Polyakov, "Gauge theory correlators from non-critical string theory," *Phys. Lett. B* **428** (1998) 105 [hep-th/9802109].

[4] J. M. Maldacena, "Wilson loops in large N field theories," *Phys. Rev. Lett.* **80** (1998) 4859-4862 [hep-th/9803002]; S.-J. Rey and J.-T. Yee, "Macroscopic strings as heavy quarks in large N gauge theory and anti-de Sitter supergravity," *Eur. Phys. J.* **C22** (2001) 379–394 [hep-th/9803001]; Y. Kinar, E. Schreiber, and J. Sonnenschein, "$Q\bar{Q}$ Potential from Strings in Curved Spacetime - Classical Results," *Nucl.Phys.* **B566** (2000) 103-125 [9811192]; J. Gomis and F. Passerini "Holographic Wilson Loops," *JHEP* **0608** (2006) 074 [hep-th/0604007].

[5] E. Witten, "Anti-de Sitter space, thermal phase transition, and confinement in gauge theories," *Adv. Theor. Math. Phys.* **2** (1998) 505 [hep-th/9803131]; S.-J. Rey, S. Theisen and J.-T. Yee, "Wilson-Polyakov Loop at Finite Temperature in Large N Gauge Theory and Anti-de Sitter Supergravity," *Nucl.Phys.* **B527** (1998) 171-186 [hep-th/9803135]; A. Brandhuber, N. Itzhaki, J. Sonnenschein and S. Yankielowicz, "Wilson Loops in the Large N Limit at Finite Temperature," *Phys.Lett.* **B434** (1998) 36-40 [hep-th/9803137]; H. Boschi-Filho, N. R. F. Braga , C. N. Ferreira, "Heavy quark potential at finite temperature from gauge/string duality," *Phys. Rev* **D74** (2006) 086001 [hep-th/0607038].

[6] N. Itzhaki, J. M. Maldacena, J. Sonnenschein, S. Yankielowicz, " Supergravity and The Large N Limit of Theories With Sixteen Supercharges ," *Phys.Rev.* **D58** (1998) 046004 [hep-th/9802042]; O. Aharony, A. Fayyazuddin, J. Maldacena, "The Large N Limit of N =2,1 Field Theories from Threebranes in F-theory," *JHEP* **9807** (1998) 013 [hep-th/9806159].

[7] L. Girardello, M. Petrini, M. Porrati, A. Zaffaroni, "The Supergravity Dual of N=1 Super Yang-Mills Theory," *Nucl.Phys.* **B569** (2000) 451-469 [hep-th/9909047]; J. Polchinski and M. J. Strasslerv," The String Dual of a Confining Four-Dimensional Gauge Theory," [hep-th/0003136]; J. Babington, D. E. Crooks, N. Evans," A Stable Supergravity Dual of Non-supersymmetric Glue ," *Phys.Rev.* **D67** (2003) 066007 [hep-th/0210068]; G.V. Efimov, A.C. Kalloniatis, S.N. Nedelko, "Confining Properties of the Homogeneous Self-Dual Field and the Effective Potential in SU(2) Yang-Mills Theory," *Phys.Rev.* **D59** (1999) 014026 [hep-th/9806165].

[8] T. Mateos, J. M. Pons, P. Talavera, "Supergravity Dual of Noncommutative N=1 SYM," *Nucl.Phys.* **B651** (2003) 291-312 [hep-th/0209150]; E. G. Gimon, L. A. P. Zayas, J. Sonnenschein, M. J. Strassler, "A Soluble String Theory of Hadrons," *JHEP* **0305** (2003) 039 [hep-th/0212061].

[9] R. Casero, C. Nunez, A. Paredes, "Towards the String Dual of N=1 Supersymmetric QCD-like Theories," *Phys.Rev.* **D73** (2006) 086005 [hep-th/0602027].

[10] I. R. Klebanov and E. Witten, "Superconformal field theory on threebranes at a Calabi-Yau singularity," *Nucl. Phys.* **B536** (1998) 199 [hep-th/9807080].

[11] I. R. Klebanov and A. A. Tseytlin, "Gravity duals of supersymmetric SU(N) × SU(N+M) gauge theories," *Nucl. Phys.* **B 578** (2000) 123 [hep-th/0002159].

[12] I. R. Klebanov and M. J. Strassler, "Supergravity and a confining gauge theory: Duality cascades and χSB-resolution of naked singularities," JHEP 0008 (2000) 052 [hep-th/0007191].

[13] J. M. Maldacena and C. Núñez, "Towards the large N limit of pure N = 1 super Yang Mills," *Phys. Rev. Lett.* **86** (2001) 588 [hep-th/0008001].

[14] A. Hashimoto and N. Itzhaki,"Non-Commutative Yang-Mills and the AdS/CFT Correspondence," *Phys.Lett.* **B465** (1999) 142 [hep-th/9907166].

[15] J. M. Maldacena and J. G. Russo," Large N Limit of Non-Commutative Theories," *JHEP* **9909** (1999) 025 [hep-th/9908134]; U. H. Danielsson, A. Guijosa, M. Kruczenski, and B. Sundborg,"D3-brane Holography," *JHEP* **0005** (2000) 028 [hep-th/0004187]; S. R. Das and B. Ghosh,"A Note on Supergravity Duals of Noncommutative Yang-Mills Theory," *JHEP* **0006** (2000) 043 [hep-th/0005007].

[16] Wung-Hong Huang, "Dual String Description of Wilson Loop in Non-commutative Gauge Theory," [hep-th/0701069].

[17] A. Hashimoto and K. Thomas, "Dualities, Twists, and Gauge Theories with Non-Constant Non-Commutativity," *JHEP* **0501** (2005) 033 [hep-th/0410123]; A. Hashimoto and K. Thomas, "Non-commutative gauge theory on D-branes in Melvin Universes," *JHEP* **0601** (2006) 083 [hep-th/0511197].

[18] F. Dowker, J. P. Gauntlett, D. A. Kastor and J. Traschen, "The decay of magnetic fields in Kaluza-Klein theory," *Phys. Rev.* **D52** (1995) 6929 [hep-th/9507143]; M. S. Costa and M. Gutperle, "The Kaluza-Klein Melvin solution in M-theory," *JHEP* **0103** (2001) 027 [hep-th/0012072]; M.A. Melvin, "Pure magnetic and electric geons," *Phys. Lett.* **8** (1964) 65;

[19] C. G. Callan, J. A. Harvey and A. Strominger, "Supersymmetric string solitons," [hep-th/9112030]; A. Dabholkar, G. W. Gibbons, J. A. Harvey and F. Ruiz Ruiz, "Superstrings And Solitons," *Nucl. Phys. B* **340** (1990) 33; G.T. Horowitz and A. Strominger, "Black strings and P-branes," *Nucl. Phys. B* **360** (1991) 197.

[20] P. Ginsparg and C. Vafa, *Nucl. Phys.* **B289** (1987) 414; T. Buscher, *Phys. Lett.* **B159** (1985) 127; B194 (1987) 59; B201 (1988) 466; S. F. Hassan, "T-Duality, Space-time Spinors and R-R Fields in Curved Backgrounds," *Nucl.Phys.* **B568** (2000) 145 [hep-th/9907152].

[21] N. Seiberg and E. Witten, "String Theory and Non-commutative Geometry," *JHEP* **09** (1999) 032 [hep-th/9908142].

[22] N. Drukker and B. Fiol, "All-genus calculation of Wilson loops using D-branes," *JHEP* **02** (2005) 010 [hep-th/0501109].

[23] S. A. Hartnoll and S. Prem Kumar, "Multiply wound Polyakov loops at strong coupling," *Phys. Rev.* **D74** (2006) 026001[hep-th/0603190]; S. Yamaguchi, "Wilson loops of anti-symmetric representation and D5-branes," *JHEP* **05** (2006) 037 [hep-th/0603208].

[24] D. J. Gross and H. Ooguri, "Aspects of large N gauge theory dynamics as seen by string theory," *Phys. Rev.* **D58** (1998) 106002 [hep-th/9805129].

[25] K. Zarembo, "Wilson loop correlator in the AdS/CFT correspondence," *Phys. Lett.* **B459** (1999) 527-534 [hep-th/9904149]; P. Olesen and K. Zarembo, "Phase transition in Wilson loop correlator from AdS/CFT correspondence," [hep-th/0009210].

[26] H. Kim, D. K. Park, S. Tamarian and H. J. W. Muller-Kirsten, "Gross-Ooguri phase transition at zero and finite temperature: Two circular Wilson loop case," *JHEP* **03** (2001) 003 [hep-th/0101235]; G. Arutyunov, J. Plefka and M. Staudacher, "Limiting geometries of two circular Maldacena-Wilson loop operators," *JHEP* **12** (2001) 014 [hep-th/0111290].

[27] A. Tsuji, "Holography of Wilson loop correlator and spinning strings," [hep-th/0606030]; C. Ahn, "Two circular Wilson loops and marginal deformations,"[hep-th/0606073]; Ta-Sheng Tai and S. Yamaguchi,"Correlator of Fundamental and Anti-symmetric Wilson Loops in AdS/CFT Correspondence," [hep-th/0610275].

In: High Energy Physics Research Advances
Editors: T.P. Harrison et al, pp. 111-135

ISBN 978-1-60456-304-7
© 2008 Nova Science Publishers, Inc.

Chapter 5

APPLICATION OF GEOMETRIC PHASE
IN QUANTUM COMPUTATIONS

A.E. Shalyt-Margolin, *V.I. Strazev and A.Ya. Tregubovich*[†]
National Centre of High Energy Physics,
220040 Bogdanovicha str. 153, Minsk, Belarus

Abstract

Geometric phase that manifests itself in number of optic and nuclear experiments is shown to be a useful tool for realization of quantum computations in so called holonomic quantum computer model (HQCM). This model is considered as an externally driven quantum system with adiabatic evolution law and finite number of the energy levels. The corresponding evolution operators represent quantum gates of HQCM. The explicit expression for the gates is derived both for one-qubit and for multi-qubit quantum gates as Abelian and non-Abelian geometric phases provided the energy levels to be time-independent or in other words for rotational adiabatic evolution of the system. Application of non-adiabatic geometric-like phases in quantum computations is also discussed for a Caldeira-Legett-type model (one-qubit gates) and for the spin 3/2 quadrupole NMR model (two-qubit gates). Generic quantum gates for these two models are derived. The possibility of construction of the universal quantum gates in both cases is shown.

Keywords: Quantum computer, Berry phase, Non-adiabatic geometric phase, Two-qubit gates

1. Introduction

The conceptions of quantum computer (QC) and quantum computation developed in 80-th [1], [2] were found to be fruitful both for computer science and mathematics as well as for physics [3]. Although a device being able to perform quantum computations is now far away from practical realization, there is a great number of theoretical proposals of such

---

[*]E-mail address: alexm@hep.by
[†]E-mail address: a.tregub@open.by

a construct (see e.g. [6]–[18]). Intensive investigations on quantum information theory (see e.g. [4], [5] for a reference source on this subject) refreshed some interest on Berry phase [56]. The idea of using unitary transformations produced by Berry phase as quantum computations is proposed in [19], [20] and first realized in [21], [47] in a concrete model of holonomic quantum computer where the degenerate states of laser beams in non-linear Kerr cell are interpreted as qubits. For other references where Abelian Berry phase is considered in the context of quantum computer see e.g. [22] - [25]. If the corresponding energy level is degenerate non-Abelian phase takes place [57] that is actually a matrix mixing the states with the same energy. For further references on quantum computation based on non-Abelian geometric phase see e.g. [47]–[53].

On the other hand non-adiabatic analogue of Berry phase can exist and be measured if transitions in a given statistical ensemble do not lead to loose of coherence [71]. For loose of coherence in quantum computations related to geometric phase see [41]–[46]. Thus it is also possible to use the corresponding unitary operators to realize quantum gates. This fact has been noticed in [26], [27]. After that a lot of papers was published where the non-adiabatic phase is applied to realize the basic gates in different models of QC such as different NMR schemes [28]–[35], ion traps [36], [37], quantum dots [38], [39], and superconducting nanocirquits [40].

To analyze a concrete scheme for quantum computation based on geometric phase it is desirable to be aware of analytical expression for the evolution operator of the system at least at the moment when the measurement is performed. This article is concentrated on the computational aspect of geometric phase for the models which are relevant to QC. It should be emphasized that the form of the expression for the phase and the possibility of the derivation of such a formula itself thoroughly depend on the group-theoretic structure of the corresponding Hamiltonian. Therefore a method of the geometric phase calculation which would be more or less universal at least in the adiabatic case can appear to be useful. The material is divided in two parts. In section 2. the adiabatic geometric phase is considered. In subsection 2.1. we analyze the difficulties appearing in calculation of the Abelian adiabatic geometric phase (Berry's phase) and propose a method of its explicit derivation for the case of the symmetric time-dependent Hamiltonian with constant non-degenerate energy levels. The symmetry of the Hamiltonian is supposed to reduce the Hamiltonian to that of a system with finite number of energy levels. In subsection 2.2. this method is generalized for the case when degeneration is present. In section 3. we consider non-adiabatic phase which can only conditionally be called "geometric" for its dependence on concrete details of the dynamics. For this reason it is not possible to work out more or less general approach to the calculation of the non-adiabatic phase. Therefore two concrete cases are considered. In subsection 3.1. application of the Abelian non-adiabatic phase to one-qubit computation in a Caldeira-Legett-type model is cosidered. In subsection 3.2. we present an example of both non-Abelian and non-adiabatic phase computation in spin-3/2 quadrupole NMR resonance model.

2. Adiabatic Geometric Phase

2.1. Abelian Berry's Phase

Here we consider a possible method of the adiabatic phase computation that seems to be effective in a broad range of practically relevant cases. Berry phase is a consequence of the adiabatic (or Born–Fock) theorem [55] which states that a parametric quantum system depending on a set of slowly (adiabatically) evolving parameters $R_i(t)$, $i = 1,\ldots N$ behaves in a quasi-stationary manner

$$\hat{H}(\boldsymbol{R})|n(\boldsymbol{R})> = E_n(\boldsymbol{R})|n(\boldsymbol{R})>, \qquad \boldsymbol{R} = (R_1,\ldots,R_N) \tag{1}$$

where $\hat{H}(\boldsymbol{R})$ is the corresponding Hamiltonian and no energy level degeneration is assumed. The adiabaticity condition means that the frequencies $\omega_n(\boldsymbol{R}) = E_n(\boldsymbol{R})/\hbar$ are much greater than the characteristic Fourier frequencies of $R_i(t)$. Thus the eigenvectors $|n(\boldsymbol{R})>$ evolve like

$$|n(\boldsymbol{R})> = \hat{S}(\boldsymbol{R})|n_0>, \qquad |n_0> = |n(\boldsymbol{R}(0))>, \quad \hat{S}\hat{S}^\dagger = 1 \tag{2}$$

with unitary rotation $\hat{S}$ describing the natural variation of $|n(\boldsymbol{R})>$ due to that of $\boldsymbol{R}(t)$. It corresponds to the following evolution law of the Hamiltonian

$$\hat{H}(t) = \hat{S}(\boldsymbol{R})\,\hat{H}_0(t)\,\hat{S}^\dagger(\boldsymbol{R}) \tag{3}$$

where $H_0(t)$ is diagonal in the basis $\{|n_0>\}$. What is the solution of the non-stationary Schrödinger equation

$$i\hbar \frac{\partial}{\partial t}|\psi(t)> = \hat{H}(t)|\psi(t)> \tag{4}$$

for this case? A natural hypothesis would be that the evolution operator for $|\psi>$

$$|\psi(t)> = \hat{U}(t)|\psi(0)>$$

has the form

$$\hat{U}(t) = \hat{S}(\boldsymbol{R})\hat{\Phi}(t)$$

where $\hat{S}$ is defined by (2) and $\hat{\Phi}$ simply produces the dynamic phase

$$\hat{\Phi}(t)|n(\boldsymbol{R}(t)> = \exp\left(-i/\hbar \int\limits_0^t E_n(\tau)\,d\tau\right)|n(\boldsymbol{R}(t)>. \tag{5}$$

It is based on the analogue with the stationary case where evolution is simply represented by the dynamical phase factor $\exp(-i/\hbar\, E_n\, t)$. Berry first observed [56] that the hypothesis is wrong. To see this it is sufficient to represent $\hat{U}$ in the form $\hat{U}(t) = \hat{S}(\boldsymbol{R})\hat{V}(t)$ and substitute it into the Schrödinger equation

$$i\hbar \frac{\partial}{\partial t}\hat{U}(t) = \hat{H}(t)\hat{U}(t). \tag{6}$$

It gives

$$i\hbar\left(\dot{\hat{V}}\hat{V}^\dagger + \hat{S}^\dagger \nabla_{\boldsymbol{R}}\hat{S}\,\dot{\boldsymbol{R}}\right) = \hat{H}_0(t). \tag{7}$$

Now one can see that $\hat{V}$ cannot be simply $\hat{\Phi}$ because it has to cancel the second term in the right hand side of (7) besides of H_0. It follows from (7)that

$$\hat{V}(t) = \hat{\Gamma}(t)\hat{\Phi}(t)$$

where $\hat{\Phi}(t)$ is determined by (5) and the following equation is valid for $\hat{\Gamma}(t)$:

$$\dot{\hat{\Gamma}}\hat{\Gamma}^{\dagger} |n_0> = -(\hat{S}^{\dagger}\nabla_{\boldsymbol{R}}\hat{S})\dot{\boldsymbol{R}} |n_0> . \tag{8}$$

It results in the evolution law for the state vector corresponding to the n-th energy level

$$|\psi_n(t)> = e^{-i/\hbar \Phi_n(t)} e^{i\gamma_n(t)} |n(\boldsymbol{R}(t))> \tag{9}$$

where $\Phi_n(t)$ is the phase factor in the right-hand side of (5) and $\gamma_n(t)$ is given by

$$\gamma_n(t) = \int_0^t \boldsymbol{A}_n(\boldsymbol{R}(\tau))\dot{\boldsymbol{R}}(\tau)\, d\tau, \quad \boldsymbol{A}(\boldsymbol{R}) = i<n(\boldsymbol{R})|\nabla_{\boldsymbol{R}}n(\boldsymbol{R})> . \tag{10}$$

Phase $\gamma_n(t)$ becomes purely geometric while $\boldsymbol{R}(t)$ evolves cyclically: $\boldsymbol{R}(T) = \boldsymbol{R}(0)$

$$\gamma_n(T) \equiv \gamma_n(\mathcal{C}) = \oint_{\mathcal{C}} \boldsymbol{A}_n(\boldsymbol{R})\, d\boldsymbol{R}. \tag{11}$$

Here the integration contour $\mathcal{C}$ is a closed curve in the parameter space described by $\boldsymbol{R}(t)$ as a radius-vector. It is easily seen from (11) that $\gamma_n(\mathcal{C})$ does not depend on the concrete details of the system's dynamic if the adiabatic condition is held.

In this article we are interested in computing of $\gamma_n(\mathcal{C})$ in the most general case. The problem of derivation of $\gamma_n(\mathcal{C})$ was solved in various particular cases in large number of articles some years ago. First we would like to note that the straightforward formula

$$\boldsymbol{B}_n = \nabla_{\boldsymbol{R}} \times \boldsymbol{A}_n = \sum_{m \neq n} \frac{(\nabla_{\boldsymbol{R}}\hat{H}(\boldsymbol{R}))_{mn} \times (\nabla_{\boldsymbol{R}}\hat{H}(\boldsymbol{R}))_{nm}}{(E_n(\boldsymbol{R}) - (E_m(\boldsymbol{R}))^2} \tag{12}$$

derived by Berry [56] by making use of the identity

$$<m|\nabla n> = \frac{<m|\nabla\hat{H}|n>}{(E_n - E_m)}, \quad m \neq n$$

has not (despite of it's beauty) much practical use because to apply it one should establish the analytical dependence of all E_n on $\boldsymbol{R}$ that is not a realistic task excluding some special cases. To see this one should attempt to apply formula (12) to the case of a 3-level system substituting a generic solution $E_n(\boldsymbol{R})$ of the corresponding cubic equation therein.

It was first noticed in [61] that symmetries of the Hamiltonian $\hat{H}(\boldsymbol{R})$ play an important role in computing of γ_n. Indeed if one represents the result of the periodic motion $\hat{H}(0) = \hat{H}(T)$ as

$$|\psi(T)> = \hat{U}(T)|\psi(0)> \tag{13}$$

Table 1. Expressions for the operators $\hat{X}_{\pm}$, $\hat{X}_3$.

Algebra	$\hat{X}_+$	$\hat{X}_-$	$\hat{X}_3$	Commutators
$su(2)$	$\hat{J}_+$	$\hat{J}_-$	$\hat{J}_3$	$[\hat{J}_3, \hat{J}_{\pm}] = \pm\hat{J}_{\pm}$ $[\hat{J}_+, \hat{J}_-] = 2\hat{J}_3$
$su(1.1)$	$\hat{K}_+$	$\hat{K}_-$	$\hat{K}_3$	$[\hat{K}_3, \hat{K}_{\pm}] = \pm\hat{K}_{\pm}$ $[\hat{K}_+, \hat{K}_-] = -2\hat{K}_3$
$H - W$	$\hat{a}^+$	$\hat{a}$	$\hat{1}$	$[\hat{1}, \hat{a}^+] = [\hat{1}, \hat{a}] = 0$ $[\hat{a}, \hat{a}^+] = \hat{1}$

where as it follows from (1) $\hat{U}(T)$ must commute with $\hat{H}(0)$ so it can be represented as an exponent containing a linear combination of operators $\hat{X}_k$ which must commute with $\hat{H}(0)$ as well. Thus the operators $\hat{X}_k$ are integrals of motion and describe certain symmetries of the given system. Therefore in what follows we restrict ourselves with such systems whose Hamiltonian is an element of a finite Lie algebra. This assumption immediately gives the group-theoretic structure of $\hat{U}(T)$:

$$\hat{U}(T) = \exp\left(i \sum_i a_i H_i\right) \tag{14}$$

where H_i are all linearly independent elements of the Cartan subalgebra and a_i are some coefficients. Thus the problem reduces to computing of the coefficients a_i. In the simplest case of Lie algebras consisting of three elements this problem can be easily solved [62], [64], [65] for physically relevant cases of Heisenberg-Weyl algebra, $su(2)$ and $su(1.1)$. In each of them the evolution operator has the form

$$\hat{U}(t) = \exp\left(\zeta(t)\hat{X}_+ - \zeta^*(t)\hat{X}_-\right) \exp\left(i\,\phi(t)\,\hat{X}_3\right) \tag{15}$$

provided the initial Hamiltonian is proportional to $\hat{X}_3$ where $\hat{X}_{\pm}$ and $\hat{X}_3$ are the corresponding generators of the algebras above. Their expressions for each concrete case are given in table 1.

The Hamiltonian for $su(2)$ case describes an arbitrary spin in the magnetic field so all J's are the angular momentum operators: $\hat{J}_{\pm} = 1/2(\hat{J}_1 \pm \hat{J}_2)$. $su(1.1)$ case corresponds to the evolution of squeezed states [66] of light in non-linear optics. Here $\hat{K}_+ = \hat{a}^{+^2}/2$, $\hat{K}_- = \hat{a}^2/2$ and $\hat{K}_3 = \hat{a}^+\hat{a} + 1/2$ where $\hat{a}$, $\hat{a}^+$ are usual bosonic annihilation and creation operators. The last case represents a harmonic oscillator interacting with the time-dependent electric field. The simple commutation relations in these three algebras admit

Table 2. Expressions for the forms ω and $d \wedge \omega$.

Algebra	$\omega(\xi)$	$d \wedge \omega(\xi)$	Relation to ζ
$su(2)$	$\dfrac{\xi d\xi^* - \xi^* d\xi}{1 + |\xi|^2}$	$\dfrac{2\, d\xi \wedge d\xi^*}{(1 + |\xi|^2)^2}$	$|\xi| = \tan(|\zeta|),$ $\arg \xi = \arg \zeta$
$su(1.1)$	$\dfrac{\xi d\xi^* - \xi^* d\xi}{1 - |\xi|^2}$	$\dfrac{2\, d\xi \wedge d\xi^*}{(1 - |\xi|^2)^2}$	$|\xi| = \tanh(|\zeta|),$ $\arg \xi = \arg \zeta$
H-W	$\xi d\xi^* - \xi^* d\xi$	$2\, d\xi \wedge d\xi^*$	$\xi = \zeta$

direct computation of Berry's phase [62], [64], [65].

$$\gamma_m = m \oint_{\mathcal{C}} \omega(\xi) = \int_S d \wedge \omega(\xi) \tag{16}$$

where m is an eigenvalue of the corresponding $\hat{X}_3$, S is the surface in the parameter space bounded by the closed curve C and the expressions for $\omega(\xi)$ and its external derivative $d \wedge \omega(\xi)$ are given for each case in table 2.

The geometric sense of the derived phase factor is the integral curvature over the surface bounded by the contour $\mathcal{C}$ on the manifold the evolution operator belongs to. This manifold can be generally expressed in the form G/H where G is the group manifold and H is that of the stationary subgroup, i.e. the group whose Lie algebra consists of all operators commuting with $H(0)$ (in these three cases it is always $U(1)$). It is sphere in the case of $su(2)$, two-sheet hyperboloid in the case of $su(1.1)$ and plane in the case of Heisenberg-Weyl group. To complete the computation one has to establish correspondence between the complex parameter ξ and physical parameters of the Hamiltonian. Let us do that for $SU(2)$. It is worth to notice that the result is completely determined by the geometric properties of the group and does not depend on the concrete representation. For this reason one can chose the fundamental representation of $SU(2)$ to simplify the derivation. Thus we take $H(0) = \omega_B \sigma_3$ which corresponds to the initial eigenvectors $|\pm> = (1(0), 0(1))^T$ (T denotes transposition). The evolution operator generally parametrized by the spherical as

$$\begin{pmatrix} \cos\frac{\vartheta}{2} & -\sin\frac{\vartheta}{2}\, e^{-i\varphi} \\ \sin\frac{\vartheta}{2}\, e^{i\varphi} & \cos\frac{\vartheta}{2} \end{pmatrix} \tag{17}$$

rotates $H(0)$ into $H(t) = \omega_B\, \boldsymbol{n\sigma}$ where $\boldsymbol{n} = (\cos\varphi \sin\vartheta/2,\ \sin\varphi \sin\vartheta/2,\ \cos\vartheta/2)$ determines the direction of the magnetic field. On the other hand the direct computation of

$\exp(\zeta \hat{J}_+ - \zeta^* \hat{J}_-)$ gives for this representation $|\zeta| = \vartheta/2$, $\arg \zeta = \varphi + \pi$. It leads to the well known expressions for the fictitious "strength field"

$$B_\pm = \mp \frac{1}{2} \frac{R}{R^3} \tag{18}$$

where R denotes the true magnetic field vector in order not to confuse it with the fictitious one which determines the resulting Berry's phase. It should be noted that the correspondence $R \to \xi$ realizes the stereographic projection of the sphere with the coordinates ϑ, φ on the plane that points are labeled by ξ. The other two cases of $SU(1.1)$ and Heisenberg-Weyl groups can be considered in a similar manner.

Re-derivation of these simplest results has the intention to extract a universal idea of computing the geometric phase in more or less general case. For the sake of certainty let us assume the symmetry algebra of the Hamiltonian to be semisimple. It means that the generic evolution operator can be represented in the form

$$\hat{U}(t) = \prod_{\alpha \in \Delta_+} \hat{U}_\alpha(t) \tag{19}$$

where Δ_+ denotes the set of the positive roots α and each U_α is analogous to (15) (see also table 1 for $su(2)$ and $su(1.1)$ cases):

$$\hat{U}_\alpha(t) = \exp\left(\zeta_\alpha(t)\, \hat{E}_\alpha - \zeta_\alpha^*(t)\, \hat{E}_{-\alpha}\right) \tag{20}$$

where the standard notations for the Cartan basis [67]

$$[H_\beta, \hat{E}_{\pm\alpha}] = \pm\alpha(H_\beta)\, \hat{E}_{\pm\alpha} \quad [\hat{E}_\alpha, \hat{E}_{-\alpha}] = H_\alpha, \quad [\hat{E}_\alpha, \hat{E}_\beta] = N_{\alpha\beta}\, \hat{E}_{\alpha+\beta} \tag{21}$$

are used. The pairs of generators $\hat{E}_{\pm\alpha}$ are analogous for $\hat{J}_\pm$ in $su(2)$ and H_α's are that of $\hat{J}_3$. Here $\alpha(H_\beta)$ and $N_{\alpha\beta}$ are constants that can be chosen rational and integer correspondingly. Taking account of the consideration above leads to some more detailed form for one-cycle evolution operator $\hat{U}(T)$ (14)

$$\hat{U}(\mathcal{C}) = \exp\left(i \sum_{\alpha \in \Delta_+} a^\alpha(\mathcal{C}) H_\alpha\right). \tag{22}$$

Each pair $(\hat{E}_\alpha, \hat{E}_{-\alpha})$ makes besides of the trivial group-theoretic contribution H_α which produces the corresponding quantum number also a non-trivial one reflecting adiabatic dynamics of the system

$$a^\alpha(\mathcal{C}) = \oint_{\mathcal{C}} \theta^\alpha(\zeta) \tag{23}$$

where θ_α generally depends on all $\zeta_\alpha(t)$. Thus to solve the problem one has to find this dependence making use of commutation relations (21) and then establish the connection between the parameters $\zeta_\alpha(t)$ and the natural set of parameters R of the Hamiltonian. Unfortunately the hope to obtain a solution of even one of these two tasks that would be a non-trivial generalization of the above examples is not realistic. Neither the first part of the problem nor the second one could be solved in a way resulting in physically relevant explicit formulas having practical use. First the 1-forms θ_α fulfill Maurer-Cartan equations that express the quantity $\hat{U}^\dagger\, d\hat{U}$ in terms of the 1-forms ω_α and θ_i

$$\hat{U}^\dagger(\zeta)\, d\hat{U}(\zeta) = i(\omega_\alpha(\zeta)\, \hat{E}_\alpha + \theta_i(\zeta)\, H_i) \tag{24}$$

where the index i labels all linearly independent generators of the Cartan subalgebra (not all H_α are so). For commutation relations (21) these equations take the form

$$d \wedge \omega^\alpha = C^\alpha_{\beta k}\, \omega^\beta \wedge \theta^k + 1/2\, C^\alpha_{\beta\lambda}\, \omega^\beta \wedge \omega^\lambda \tag{25}$$

$$d \wedge \theta^i = 1/2\, C^i_{\beta\lambda}\, \omega^\beta \wedge \omega^\lambda \tag{26}$$

Equations (25), (26) describe the parallel transport on the coset manifold G/H. The possibility to solve them depends on the manifold's symmetry and of course is entirely determined by the structure constants $C^{\cdot}_{\cdot\cdot}$ that are built from the root vectors $\alpha(H_\beta)$ and the constants $N_{\alpha\beta}$. The general solution of this system can be constructed for very high symmetry of symmetric spaces [68] where the whole algebra can be split in two subsets X and Y such that

$$[Y,Y] \subset Y, \quad [Y,X] \subset X, \quad [X,X] \subset Y.$$

It is seen from (21) that the last condition is generally speaking not valid for our case because not all $N_{\alpha\beta}$ are zeroes. Its geometric sense is that the considered coset spaces G/H are of more general symmetry type than symmetric spaces. Thus for $G = SU(n)$ the space

$$SU(n)/\underbrace{U(1) \times U(1)\ldots \times U(1)}_{n-1\text{ times}} \tag{27}$$

belongs to the more general class of Kählerian spaces. The general solution of (25), (26) for the types of spaces we are interested in is not obtained so far. Therefore the practical use of these equations is not high. Moreover the solution of the second part of the problem discussed is not possible for the same reason.

A simple and effective method of practical computation of geometric phase where it is not necessary to find the forms ω^α, θ^i is proposed in [75]. For this purpose we have to make some assumptions. First we regard the Hamiltonian to belong to a finite irreducible representation of a semisimple Lie algebra therefore $\hat{H}_0$ in (3) can always be represented as a finite matrix $\hat{H}_0 = R_i(t)\, H_i$ where the set $\{H_i\}$ is a basis of the Cartan subalgebra and $R_i(t)$ are parameters. Then we suppose the energy levels E_m to be constants. It corresponds to a rotation-type evolution (3) where $\hat{H}_0$ does not depend on t. Such a situation takes place practically in all experiments on the geometric phase measurement. This makes it possible to regard E_m as additional secondary parameters to be found just once (may be numerically). The third assumption is that the spectrum remains always non-degenerate i.e., no crossing of energy levels occurs. As the spectrum of the Hamiltonian is finite, the state vector $|\varphi_m\rangle$ is a unit vector m in $\mathbf{C}^n$, so A_m is

$$A_m = \frac{i}{2}\, (m^* dm - m\, dm^*). \tag{28}$$

As the evolution is adiabatic, the spectrum of $H(t)$ remains always non-degenerate if it was so at the initial time. Then there is always a nonzero main minor of $H - E_m$ which we assume to consist always of the first $n - 1$ lines and columns of $H - E_m$. Denoting the matrix consisting of the first $n - 1$ lines and columns of H by $H_\perp$ we come to the condition

$$\det(H_\perp - E_m) \neq 0 \qquad (29)$$

Making use of this condition one can represent n in the uniform coordinates

$$m = \frac{(\boldsymbol{\xi}_m, 1)}{\sqrt{1 + |\boldsymbol{\xi}_m|^2}}$$

and express $\boldsymbol{\xi_m}$ in terms of H_{ij} for $1 \leq i, j \leq n - 1$ and E_m:

$$\boldsymbol{\xi}_m = (H_\perp - E_m)^{-1} \boldsymbol{h}, \qquad h_i = -H_{in}, \qquad (30)$$

where $\boldsymbol{h}$ is a vector in $\mathbf{C}^{n-1}$ but not in $\mathbf{C}^n$. Thus we have for A_m

$$A_m = \frac{i}{2} \frac{(\boldsymbol{\xi_m}^* d\boldsymbol{\xi_m} - \boldsymbol{\xi_m} d\boldsymbol{\xi_m}^*).}{1 + |\boldsymbol{\xi_m}|^2}, \qquad (31)$$

where $\boldsymbol{\xi}_m$ is completely determined by (30). Note that the result obtained is purely geometrical because it can be expressed of the Kählerian potential

$$F = \log(1 + |z|^2)$$

where z is a vector in $\mathbf{C}^{N^2}$ consisting of $n(n - 1)/2$ independent components of all $\boldsymbol{\xi}_m$. The function $F(z, z^*)$ determines all the geometrical properties of the state space (27). Particularly its metric tensor is

$$g_{ij} = \frac{\partial^2 F(z, z^*)}{\partial z_i \partial z_j^*}.$$

It should be emphasized that the simplification of the problem reached here is based on the fact that $\dot{E}_m = 0$ so one can include it in new parameters and use them rather than $\boldsymbol{R}_i$. Therefore the dependence of E_m on $\boldsymbol{R}$ is not required. One can calculate E_m numerically and substitute it into the formulas regarding this quantity as one more external parameter. Moreover to find γ_m one needs only the energy E_m but not the whole spectrum as in (12). It can become an important issue if one considers partially solvable models. The requirement $\dot{E}_m = 0$ is sufficient because otherwise one has to solve the secular equation at each moment t that is equivalent to the numerical solution of the non-stationary Schrödinger equation itself and therefore it makes the discussed method useless.

Let us now consider some simple applications of the proposed method. First let us see how it works for the trivial case $n = 2$. (1) reduces then to two linearly dependent equations

$$(B_3 \mp B)\xi + (B_1 - iB_2) = 0$$
$$(B_1 + iB_2)\xi + (-B_3 \mp B) = 0$$

Here we returned to the usual notations of the magnetic field components B_i and $\pm B = \pm|\boldsymbol{B}|$ is the energy of the state $|\pm>$. Choosing one of the equations and taking the spherical coordinates we come to one of the relations

$$\xi = -\tan \vartheta/2 \, e^{-i\varphi} \quad \xi = \cot \vartheta/2 \, e^{-i\varphi}.$$

for the upper and lower sign correspondingly. Thus these are the coordinates of stereographic projection made from the north (south) pole of the sphere. Substituting it into the formula for $\omega_\pm(\xi)$ (see Table 2) and integrating over a contour C we get the well known result [56]

$$\gamma_\pm = \pm\frac{1}{2} \oint\limits_{C} \frac{\xi^* d\xi - \xi d\xi^*}{1 + |\xi|^2} = \mp\frac{1}{2} \; \Omega(C),$$

where $\Omega(C)$ is the solid angle corresponding to the closed contour C on the sphere.

One more example which is less trivial is a generic three-level system. The k-th eigenvector ξ_k is then two-dimensional and some trivial algebra gives for its components

$$
\begin{aligned}
\xi_1 &= \Delta_1/\Delta_0, \quad \xi_2 = \Delta_2/\Delta_0, \\[1ex]
\Delta_0 &= (H_{11} - E_k)(H_{22} - E_k) - |H_{12}|^2 \\[1ex]
\Delta_1 &= H_{23}H_{12} - H_{13}(H_{22} - E_k) \\[1ex]
\Delta_2 &= H_{13}H_{12}^* - H_{23}(H_{11} - E_k)
\end{aligned}
\tag{32}
$$

Here we have omitted where possible the index k. Substitution of these expressions into (31) gives the final formula for this case. Note that the use of formula (12) here would lead to sufficient computational difficulties even after making further simplifying assumptions [69]. The proposed approach makes concrete calculations visibly easier and more compact although the final formulas are not of esthetic value. For the illustrative purpose we take here the case when all H_{ij}, $i \le j$ but H_{12} do not depend on t. Then substitution of (32) into (31) gives

$$\omega_k(\xi) = i C_k \left[A - D_k \sin(\phi_{12} + \phi_{23} - \phi_{13}) \right] d\phi_{12}, \tag{33}$$

$$C_k = \frac{|H_{13}||H_{23}||H_{12}|}{\Delta_0^2(E_k) + |\Delta_1(E_k)|^2 + |\Delta_2(E_k)|^2}, \tag{34}$$

$$A = 1/|H_{13}|^2 - 1/|H_{23}|^2, \tag{35}$$

$$D_k = H_{11} + H_{22} - 2E_k \tag{36}$$

where ϕ_{ij} are arguments of the complex numbers H_{ij}. As it was discussed above the condition $\Delta_0(t) \ne 0$ is supposed to be held everywhere on C. For other cases of the geometric phase in the 3-level system see [63].

2.2.   Non-abelian Wilczek–Zee Phase

Now we proceed with a more general case of degenerate spectrum. Quantum computation for this case generated by an adiabatic loop in the control manifold is determined by the same quasi-stationary Schrödinger equation (1) where each energy level E_m corresponds to a set of eigenstates $|m_a\rangle$, $a = 1, \ldots d_m$. Cyclic evolution of the parameters results in

$$|m_a(T)\rangle = U_{ab}(T) |m_b(0)\rangle \tag{37}$$

where the matrix U is presented by a $\mathcal{P}$-ordered exponent

$$U(\mathcal{C}) = \mathcal{P} \exp \left(\oint_{\mathcal{C}} \boldsymbol{A}_m \right), \qquad (\boldsymbol{A}_m)_{ab} = i <m_b \,|\, dm_a> . \qquad (38)$$

In this section we generalize the proposed approach to the geometric phase computation for the generic case of degenerate energy levels [75].

The set of eigenvectors $\boldsymbol{\xi}_{ma}$, $a = 1, ..., d_m$ must obey the equation

$$(H_\perp^{(d_m)} - E_m) \, \boldsymbol{\xi}_{ma} = h \, \boldsymbol{c}_a. \qquad (39)$$

Here the matrix $H_\perp^{(d_m)}$ is constructed from the first $n - d_m$ lines and columns of H, $\boldsymbol{c}_a$ are arbitrary d_m-dimensional vectors and h is the following $(n - d_m) \times d_m$-matrix:

$$h = - \begin{pmatrix} H_{1,n-d_m+1} & \cdots & H_{1,n} \\ \vdots & \cdots & \cdots \\ H_{n-d_m,n-d_m+1} & \cdots & H_{n-d_m,n} \end{pmatrix}$$

Of course it has sense only if the condition

$$\det(H_\perp^{(d_m)}(t) - E_m) \neq 0 \qquad (40)$$

is valid along the evolution process. The set of vectors $\boldsymbol{\xi}_{ma}$ must be orthogonalized by the standard Gram algorithm and after that we get the orthonormal set of the eigenvectors $\boldsymbol{z}_a$ (here and below we has omitted the index m) in the form

$$z_a = \frac{1}{\det \Gamma_{a-1}} \begin{pmatrix} & & & x_1 \\ & \Gamma_{a-1} & & \vdots \\ \langle \boldsymbol{\xi}_a | \boldsymbol{\xi}_1 \rangle & \cdots & \langle \boldsymbol{\xi}_a | \boldsymbol{\xi}_{a-1} \rangle & x_a \end{pmatrix}, \qquad (41)$$

where $x_b = (\boldsymbol{\xi}_b, \boldsymbol{c}_b)$ and $\boldsymbol{c}_a$ is chosen to be the standard orthogonal set $\boldsymbol{c}_a = (0... \overset{a}{\overset{\frown}{1}} ...0)$. The matrices Γ_a are determined by

$$\Gamma_a = \begin{pmatrix} 1 + \langle \boldsymbol{\xi}_1 | \boldsymbol{\xi}_1 \rangle & \cdots & \langle \boldsymbol{\xi}_1 | \boldsymbol{\xi}_a \rangle \\ \vdots & \ddots & \vdots \\ \langle \boldsymbol{\xi}_a | \boldsymbol{\xi}_1 \rangle & \cdots & 1 + \langle \boldsymbol{\xi}_a | \boldsymbol{\xi}_a \rangle \end{pmatrix} = 1 + Z_a^\dagger Z_a, \qquad (42)$$

where the $(n - d_m) \times a$-matrix Z_a consists of a first lines of the $(n - d_m) \times d_m$-matrix $Z = (H_\perp^{(d_m)} - E_m)^{-1} h$. Using (42) and (41) we come to the final expression for the matrix-valued 1-form $\boldsymbol{A}$:

$$\boldsymbol{A} = \frac{i}{2} \frac{g_{ab}^{ij}(\boldsymbol{\xi}_j^* d\boldsymbol{\xi}_i - d\boldsymbol{\xi}_j^* \boldsymbol{\xi}_i) + 2\omega_{ab}}{\det(1 + Z_{a-1}^\dagger Z_{a-1}) \det(1 + Z_{b-1}^\dagger Z_{b-i})}, \qquad 1 \leq i \leq a, \; 1 \leq j \leq b, \qquad (43)$$

where

$$g_{ab}^{ij} = \Gamma_a^i \Gamma_b^{*j}, \qquad \omega_{ab} = \langle \boldsymbol{\xi}_j \left| d\,\mathrm{Im}(g_{ab}^{ij}) \right| \boldsymbol{\xi}_i \rangle + \sum_{i=1}^{min(a,b)} d\,\mathrm{Im}(g_{ab}^{ii}),$$

and Γ_a^i is the cofactor of ξ_i in Γ_a. Note that the change of our basis c_a by $c_a' = U_{ab}(\lambda)\,c_b$ leads to a standard gauge transformation of A

$$A' = UAU^\dagger + i(dU)U^\dagger.$$

The formula (43) is the desired expression of A in terms of the matrix elements of the Hamiltonian. It is correct if condition (40) is valid. It is not nevertheless a principal restriction because d_m does not depend on time due to adiabaticity of the evolution and there is always at least one nonzero $n - d_m$-order minor of H. Then, if the minor we choose vanishes somewhere on the loop C one can always take local coordinates such that the techniques considered is applicable on each segment of C. It should also be noted at the end of this section that the idea of physical realization of the quantum gates based on the concrete system driven by external electromagnetic fields appears if one takes into account that for A_n E_α can be realized by means of ordinary bosonic creation and annihilation operators, namely $E_\alpha = a_i\dagger a_j$ for some $1 \le i, j \le n$. Then E_α represents nothing but two-mode squeezing operator. Thus the model considered can be applied to optical HQC with n laser beams (the case n=2 is considered in [21]) and the logical gates U_α are just two-qubit transformations realized by transformation of two laser beams.

The method presented here enables one to build in principal any computation for HQC described by a Hamiltonian with a stationary spectrum in terms of experimentally measured values exactly the matrix elements of the Hamiltonian. The method depends weakly on the dimension of the qubit space which other models based on various parameterizations of the system's evolution operator are very sensitive to. Application of this method to a concrete physical model will be discussed elsewhere.

3.   Non-adiabatic Geometric Phase

3.1.   Abelian Non-adiabatic Phase

The adiabatic condition of quantum system's evolution is strong enough to restrict sufficiently the scope of the search for realistic candidates for practical realization of quantum computations despite of some attractive features of the adiabatic case such as fault tolerance due to independence of the evolution law on the details of the parameters' dynamics etc. Therefore it is desirable to find physically relevant cases for which on the one hand this condition would be not necessary but on the other hand the coherency in such a system would be not yet violated so that the notion of the phase shift itself could have physical sense. As the adiabatic theorem is no longer valid the property of universality of the system's dynamics (independence of the concrete form of the functions $R_i(t)$) is no more preserved and the evolution law is sufficiently more complicated. Then one cannot hope to carry out a general approach to derivation of the corresponding phase shift because in each case it depends on the fine details of the parameters variation. For the same reason the phase can be called "geometric" only conditionally because geometric intuition is no more helpful for this case e.g. the result can be represented as an integral over t rather than over a contour that expresses mathematically the thesis above. On the other hand non-adiabatic conditional geometric phase that was theoretically predicted in [60] can be measured if transitions taking place in the system do not lead to decoherence [71]. Therefore it is also possible to use the corresponding unitary operators to perform quantum calculations. This fact has been noticed in [26], [27] (see also [28]–[35] for further references).

Let us consider a parametric quantum system described by the Hamiltonian $H(\boldsymbol{R})$, where $\boldsymbol{R}(t)$ is a set of arbitrarily evolving parameters. We suppose that evolution of the Hamiltonian is determined by unitary rotation (3) Looking for particular solutions of the Schrödinger equation (4) (here we supposed $\hbar = 1$) we take a rotating frame by assigning $\tilde{\psi}(t) = U(t)\,\psi(t)$ and get in such a way

$$i\frac{\partial\tilde{\psi}}{\partial t} = \left(H_0(t) - i\,U^{\dagger}(t)\dot{U}(t)\right)\tilde{\psi}(t). \qquad (44)$$

Of course, this transformation generally does not help to solve equation (4) due to the fact that the algebraic structure of the coupling term $-i\,U^{\dagger}\dot{U}$ can appear to be rather complicated and the last generally does not commute with H_0. However if a receipt is known how to evaluate the last term in (44), further solution of this equation is straightforward:

$$\psi(t) = e^{-i\,\phi_n(t)}\,\mathcal{T}\exp\left(-i\int_0^t U^{\dagger}(\tau)\dot{U}(\tau)\,d\tau\right)\psi(0), \qquad (45)$$

where $\phi_n(t) = \int_0^t E_n(\tau)\,d\tau$ is so called dynamic phase, $\mathcal{T}$ denotes time-ordering and E_n are elements of H_0 that is by definition diagonal. Of course if there is no way to find $U^{\dagger}(\tau)\dot{U}(\tau)$, expression (45) is useless.

Let us illustrate it for the simplest case of spin $1/2$ in the non-adiabatically rotating magnetic field [72]. Uniform rotation in the plane $\vartheta = const$ is represented by

$$H(t) = e^{\pm i\omega_R t\,\hat{J}_3}\,e^{-i\vartheta\,\hat{J}_2}\,(\Omega\,\hat{J}_3)\,e^{-i\vartheta\,\hat{J}_2}\,e^{\mp i\omega_R t\,\hat{J}_3} \qquad (46)$$

where the sign $+\,(-)$ corresponds to the left (right) polarization. Application of (44) to (46) gives for the Hamiltonian in the rotating frame

$$H_1(t) = e^{-i\vartheta\,\hat{J}_2}\,(\Omega\,\hat{J}_3)\,e^{-i\vartheta\,\hat{J}_2} \pm \omega_R\,\hat{J}_3. \qquad (47)$$

To diagonalize Hamiltonian (47) one has to apply one more rotation to it

$$H_2(t) = V\,H_1(t)\,V^{\dagger}, \quad V = e^{i\vartheta^*\,\hat{J}_2}$$

where the angle ϑ^* does not coincide with ϑ due to the second non-adiabatic term. It should rather fulfill the condition

$$\tan\vartheta^* = \frac{\sin\vartheta}{\cos\vartheta \pm \omega_R/\Omega}. \qquad (48)$$

The second term in the denominator of (48) is the measure of non-adiabaticity of the motion. It is clear that the angle ϑ^* replaces the usual azimuthal angle ϑ in the formula for the geometric-like phase:

$$\gamma_{\pm} = \mp m_3\,2\pi\,(1 - \cos\vartheta^*) \qquad (49)$$

where m_3 is the third spin projection. Formula (49) is a natural generalization of the usual Berry's formula for the adiabatic case and coincides with it in the limit $\omega_R/\Omega \to 0$. Note that the dependence of the result on ω_R reflects the fact the phase is no longer truly geometric because ω_R characterizes the rotation velocity and thus the velocity of the motion along the contour in the parameter space.

As an example of the application of the non-adiabatic formula above we propose a realization of quantum gates for a concrete 4-level quantum system driven by external magnetic field [77]. Let us consider a system of two qubits in a bosonic environment described by the Hamiltonian

$$H = H_S + H_B + H_{SB}, \tag{50}$$

where H_S is the Hamiltonian of two coupled spins

$$H_S = H_S^{(0)} + H_S^{\text{int}} = \frac{\omega_{01}}{2}\,\sigma_{z1} \otimes 1_2 + \frac{\omega_{02}}{2}\,1_2 \otimes \sigma_{z2} + \frac{J}{4}\,\sigma_{z1} \otimes \sigma_{z2}, \tag{51}$$

where J is the coupling constant, H_B is the Hamiltonian of the bosonic environment

$$H_B = \sum_k \omega_{bk}(\hat{b}_k^+ \hat{b}_k + 1/2), \tag{52}$$

and H_{SB} is the Hamiltonian of the spin- enviroment interaction.

$$H_{SB} = H_{SB}^{(1)} + H_{SB}^{(2)}, \tag{53}$$

$$H_{SB}^{(a)} = S_z^{(a)} \sum_k (g_{ak}\hat{b}_k^+ + g_{ak}^*\hat{b}_k) \quad a = 1, 2. \tag{54}$$

Here

$$S_z^{(1)} = \sigma_{z1} \otimes 1_2, \quad S_z^{(2)} = 1_2 \otimes \sigma_{z2},$$

σ_z is the third Pauli matrix, 1_2 is 2×2 unit matrix, $\hat{b}_k^+, \hat{b}_k$ are bosonic creation and annihilation operators and g_{ak} are complex constants. We assume that the two spins under consideration are not identical so that $\omega_{01} \neq \omega_{02}$. The Hamiltonian determined by (50) – (54) is a natural generalization of Caldeira-Legett Hamiltonian [74] for the case of two non-interacting spins. Let such a system be placed in the magnetic field affecting the spins but not the phonon modes. The only change to be made in the spin part (51) is the substitution

$$\omega_s\sigma_z \longrightarrow \boldsymbol{B}\boldsymbol{\sigma},$$

Three components of $\boldsymbol{B}$ represent a control set for the qubits under consideration. Evolution of $\boldsymbol{B}(t)$ generates evolution of the reduced density matrix $\rho_s(t)$ that describes the spin dynamics

$$i\frac{\partial \rho_s(t)}{\partial t} = H_S\,\rho_s(t), \quad \rho_s(t) = U(t)\rho(0)U^+(t). \tag{55}$$

Thus given curve in the control space corresponds to a quantum calculation in which each qubit is to be processed independently. To obtain such a calculation as a function of control parameters we first recall some common issues of spin dynamics. We consider the external magnetic field as a superposition of a constant component and a circular polarized wave:

$$\boldsymbol{B} = \boldsymbol{B}_0 + \boldsymbol{B}_1 e^{i\omega_R t}, \tag{56}$$

where B_0 is perpendicular to B_1. It is well known that the case of the circular polarization is exactly solvable. The evolution of an individual spin corresponding to the Hamiltonian

$$H = -\mu B \tag{57}$$

is determined by (15) where $\hat{X}_{\pm}$, $\hat{X}_3$ are replaced by $S_{\pm} = S_x \pm iS_y$, S_z correspondingly and

$$\zeta(t) = |\zeta(t)| \exp(i\Delta\omega t + i\alpha(t) + i\pi/2), \tag{58}$$

$$|\zeta(t)| = \frac{\omega_{\perp}\sin(\Omega t/2)}{\sqrt{(\Delta\omega)^2 + \omega_{\perp}^2}},$$

$$\alpha(t) = \arctan\left(\frac{\Delta\omega}{\Omega}\tan(\Omega t/2)\right),$$

$$\phi(t) = -\omega_{\perp}(\xi_1 n_2 + \xi_2 n_1), \tag{59}$$

where $\Delta\omega = \omega_{\parallel} - \omega_R$, $\Omega^2 = (\Delta\omega)^2 + \omega_{\perp}^2$, $\omega_{\perp}$ and $\omega_{\parallel}$ are Rabi frequencies corresponding to B_0 and B_1 respectively and finally n is the unit vector along B_1.

It is known [71] that the pure states acquire within the rotating wave approximation a phase factor that after one complete cycle $T = 2\pi/\omega_R$ is:

$$|m(T)\rangle = \exp(-i\phi_D + i\gamma)|m(0)\rangle, \tag{60}$$

where m is the azimuthal quantum number and the phase is split in two parts: dynamic

$$\phi_D = 2\pi m\,\frac{\Omega}{\omega_R}\,\cos(\theta - \theta^*)$$

and geometrical

$$\gamma = -2\pi m\cos\theta^*, \tag{61}$$

where $\cos\theta = B_0/B$ ($B = B_0 + B_1$) and θ^* is determined by formula (49). The phase shift between the states $|\pm 1/2\rangle$ results then in

$$\Delta\phi_g = -2\pi\cos\theta^* \tag{62}$$

that is nothing but the solid angle enclosed by the closed curve $B(0) = B(T)$ on the Bloch sphere. If the rotation is slow such that $\omega_R/\Omega \to 0$ then $\theta^* \to \theta$ and phase shift(62) coincides with the usual Berry phase.

Thus the adiabaticity condition is not really necessary for obtaining of the geometrical phase in an ensemble of spins if the decoherence time is much greater than T. Therefore one can attempt to use this phase to get quantum gates such as CNOT. Calculation of the corresponding phase factors is rather straightforward because the free and the coupling parts of the spin Hamiltonian commute with each other

$$\left[H_S^{(0)}, H_S^{\text{int}}\right] = 0.$$

Therefore the coupling part can be diagonalized simultaneously with the free part by applying of the transformation $U = U_1 \otimes U_2$ where $U_{1,2}$ are the diagonalizing matrices for each single-spin Hamiltonian respectively. This simple fact together with the following obvious identity

$$U^\dagger \dot{U} = U_1^\dagger \dot{U}_1 \otimes 1_2 + 1_2 \otimes U_2^\dagger \dot{U}_2$$

the final formula for the part of the evolution operator that stands for the non-adiabatic geometric phase

$$U_g = \exp(-2\pi i \cos\theta_1^* \, S_{1z}) \otimes \exp(-2\pi i \cos\theta_2^* \, S_{2z}), \tag{63}$$

where

$$\tan\theta_1^* = \frac{\sin\theta_1}{\cos\theta_1 + \omega_R/\Omega_1}, \quad \tan\theta_2^* = \frac{\sin\theta_2}{\cos\theta_2 + \omega_R/\Omega_2}$$

and

$$\cos\theta_1 = \omega_{01}/\Omega_1, \quad \Omega_1^2 = \omega_{01}^2 + \omega_1^2,$$

$$\cos\theta_2 = \omega_{02}/\Omega_2, \quad \Omega_2^2 = \omega_{02}^2 + \omega_1^2.$$

Note that gate (63) is symmetric with respect to the spin transposition as it should be and does not depend on J that is typical for geometrical phase in spin systems where the phase depends only on the position drawn by the vector B on the Bloch sphere. As J does not affect this position, it is absent in the final result. We do not consider here the dynamic phase determining by the factor

$$U_d = \exp\left(-\frac{i}{\hbar}\hat{H}_S T\right).$$

It is so because one can eliminate it by making use of the net effect of the compound transformation proposed in [22]. After this transformation that is generated by two different specifically chosen contours the dynamic phase acquired by the different spin states becomes the same and the geometric phase of each state is counted twice. After that we get (up to a global phase) the following quantum gate

$$U_g = \begin{pmatrix} e^{i(\gamma_1+\gamma_2)} & 0 & 0 & 0 \\ 0 & e^{i(\gamma_1-\gamma_2)} & 0 & 0 \\ 0 & 0 & e^{i(-\gamma_1+\gamma_2)} & 0 \\ 0 & 0 & 0 & e^{-i(\gamma_1+\gamma_2)} \end{pmatrix}. \tag{64}$$

Thus we have constructed the quantum gate, which possess the advantage to be fault tolerant with respect to some kinds of errors such as the error of the amplitude control of B. On the other hand this approach makes it possible to get rid of the adiabaticity condition that strongly restricts the applicability of the gate. Instead of this condition one needs some more weak one: $\tau \gg \omega_R^{-1}$, where τ is the decoherence time.

3.2.   Non-abelian and Non-adiabatic Phase

In this section we give an example of both non-Abelian and non-adiabatic phase for a concrete 4-level quantum system driven by external magnetic field [76]. Let us consider a spin-3/2 system with quadrupole interaction. Physically it can be thought of as a single spin-3/2 nucleus. A coherent ensemble of such nuclei manifest geometric phase when placed in rotating magnetic field. This phase is non-Abelian due to degenerate energy levels with respect to the sign of the spin projection. Depending on the experiment setup the phase can be both adiabatic as in Rb experiment by Tycko [70] and non-adiabatic as in Xe experiment by Appelt et al [71]. This non- Abelian phase results in mixing of $\pm 1/2$ states in one subspace and $\pm 3/2$ in another one and thus can be regarded as a 2-qubit gate. The gate is generated by a non-Abelian effective gauge potential $\mathbf{A}$ that is the subject of computation in this section.

We assume the condition of the ^{131}Xe NMR experiment to be held so one does not need to trouble about the coherency in the system. The last is described by the following Hamiltonian in the frame where the magnetic field is parallel to the z-axis ($\hbar = 1$)

$$H_0 = \omega_0(J_3^2 - 1/3j(j+1)). \tag{65}$$

Here and in what follows we omitted the hat symbol over all J's for the sake of simplicity. For a spin-3/2 system we choose the third projection of the angular momentum in the form

$$J_3 = \begin{pmatrix} 3/2 & 0 & 0 & 0 \\ 0 & -3/2 & 0 & 0 \\ 0 & 0 & 1/2 & 0 \\ 0 & 0 & 0 & -1/2 \end{pmatrix} = \begin{pmatrix} 3/2\sigma_3 & 0 \\ 0 & 1/2\sigma_3 \end{pmatrix}, \tag{66}$$

Then two other projection operators are

$$J_1 = \begin{pmatrix} 0 & 0 & \sqrt{3}/2 & 0 \\ 0 & 0 & 0 & \sqrt{3}/2 \\ \sqrt{3}/2 & 0 & 0 & 1 \\ 0 & \sqrt{3}/2 & 1 & 0 \end{pmatrix} = \begin{pmatrix} 0 & \frac{\sqrt{3}}{2} \\ \frac{\sqrt{3}}{2} & \sigma_1 \end{pmatrix}, \tag{67}$$

$$J_2 = \begin{pmatrix} 0 & 0 & 0 & -\sqrt{3}/2 \\ 0 & 0 & \sqrt{3}/2 & 0 \\ 0 & \sqrt{3}/2 & 0 & -i \\ -\sqrt{3}/2 & 0 & i & 0 \end{pmatrix} = \begin{pmatrix} 0 & -i\frac{\sqrt{3}}{2}\sigma_3 \\ i\frac{\sqrt{3}}{2}\sigma_3 & \sigma_2 \end{pmatrix}. \tag{68}$$

In the laboratory frame the Hamiltonian takes the form

$$H = \omega_0((\boldsymbol{Jn})^2 - 1/3j(j+1)) = e^{-i\varphi J_3}e^{-i\theta J_2} H_0 e^{i\varphi J_2}e^{i\theta J_3}. \tag{69}$$

Rotation around the z-axis means that $\varphi = \omega_1 t$ and one should perform the unitary transformation

$$|\psi> = U_1|\tilde{\psi}>, \qquad U_1 = e^{-i\omega_1 t J_3}. \tag{70}$$

In the rotating frame we get

$$H_1 = e^{-i\theta J_2}(\omega_0 J_3^2 - \omega_1 \tilde{J}_3)e^{i\theta J_2} - \frac{5\omega_0}{4}, \tag{71}$$

where

$$\tilde{J}_3 = e^{i\theta J_2} J_3 e^{-i\theta J_2}$$

Expression (71) is equivalent to

$$H_1 = \begin{pmatrix} \omega_0 - \frac{3}{2}\omega_1 \cos\theta\, \sigma_3 & \omega_1 \sqrt{3}/2 \\ \omega_1 \sqrt{3}/2 & -\omega_0 - \frac{1}{2}\omega_1 \cos\theta\, \sigma_3 + \omega_1 \sin\theta\, \sigma_1 \end{pmatrix} \tag{72}$$

It is convenient to diagonalize this matrix in two steps. First we get rid of σ_1 in the last matrix element by performing of the block-diagonal transformation

$$U_2 = \mathrm{diag}(1, e^{-i\alpha\sigma_3}), \tag{73}$$

where $\tan\alpha = 2\tan\theta$. Thereafter the Hamiltonian H_1 reads

$$H_1 = \begin{pmatrix} \omega_0 - \frac{3}{2}\,\omega_1\,\cos\theta\,\sigma_3 & \omega_1\,\sqrt{3}/2 \\ \omega_1\,\sqrt{3}/2 & -\omega_0 - \frac{1}{2}\,\omega_1\,\frac{\cos\theta}{\cos\alpha}\,\sigma_3 \end{pmatrix}. \tag{74}$$

At the second step we apply the transformation

$$\begin{pmatrix} \beta_1 & \beta_2 \\ -\beta_2^* & \beta_1^* \end{pmatrix}, \tag{75}$$

where β_1, β_2 are diagonal 2×2 matrices that must obey the unitarity condition

$$|\beta_1|^2 + |\beta_2|^2 = 1. \tag{76}$$

Supposing $\beta_{1,2}$ to be real and performing transformation (75) we come to the diagonalization condition in the form

$$\xi(\beta_1^2 - \beta_2^2) + (\lambda_1 - \lambda_2)\beta_1\beta_2 = 0, \tag{77}$$

where λ_1, λ_2 are 2×2 diagonal matrices and ξ is a parameter

$$\lambda_1 = \omega_0 - 3/2\,\omega_1 \cos\theta\,\sigma_3, \tag{78}$$

$$\lambda_2 = -\omega_0 - 1/2\,\omega_1\frac{\cos\theta}{\cos\alpha}\,\sigma_3 \tag{79}$$

$$\xi = \omega_1\sqrt{3}/2 \sin\theta \tag{80}$$

Assuming $\beta_2 = \mu\beta_1$ where μ is a diagonal 2×2 matrix as well we come to the following expressions for the matrix elements of μ

$$\mu_i = k_i + \sqrt{1 + k_i^2}, \tag{81}$$

where

$$k_i = \frac{\Delta\lambda_i}{2\xi}, \quad \Delta\lambda_i = \lambda_{1i} - \lambda_{2i}.$$

Finally we get for the matrix elements of $\beta_{1,2}$

$$\beta_{1i}^2 = 1/2(1+k_i^2)^{-1/2}\left(k_i + \sqrt{1+k_i^2}\right)^{-1} \tag{82}$$

$$\beta_{2i}^2 = 1/2\left(1 + \frac{k_i}{\sqrt{1+k_i^2}}\right) \tag{83}$$

Now one can evaluate the connection 1-form. It is convenient to represent it as follows:

$$\boldsymbol{A} = i\,U^+dU = A\,d\phi = \begin{pmatrix} A_{3/2} & A^{tr} \\ \tilde{A}^{tr} & A_{1/2} \end{pmatrix} d\phi, \tag{84}$$

where all matrix elements of A denote 2×2 matrix-valued blocks, $U = U_1U_2U_3$ and U_i are determined by (70), (73), (75) correspondingly. Here tilde denotes a transposed matrix. After some algebra we get for the matrix elements of (84)

$$A^{tr} = \frac{1}{2}\beta_1\beta_2(3-\cos\alpha)\sigma_3 + \frac{1}{2}\sin\alpha\,\beta_2\sigma_1\beta_1, \tag{85}$$

$$A_{3/2} = (a_{3/2} + b_{3/2}\,\sigma_3 + c_{3/2}\,\sigma_1)\,d\phi, \tag{86}$$

$$a_{3/2} = \frac{1}{4}\left(3\beta_{11}^2 - 3\beta_{12}^2 + \beta_{21}^2\cos\alpha - \beta_{22}^2\cos\alpha\right), \tag{87}$$

$$b_{3/2} = \frac{1}{4}\left(3\beta_{11}^2 + 3\beta_{12}^2 + \beta_{21}^2\cos\alpha + \beta_{22}^2\cos\alpha\right), \tag{88}$$

$$c_{3/2} = -\frac{1}{2}\sin\alpha\,\beta_{21}\beta_{22}, \tag{89}$$

$$A_{1/2} = (a_{1/2} + b_{1/2}\,\sigma_3 + c_{1/2}\,\sigma_1)\,d\phi, \tag{90}$$

$$a_{1/2} = \frac{1}{4}\left(3\beta_{21}^2 - 3\beta_{22}^2 + \beta_{11}^2\cos\alpha - \beta_{12}^2\cos\alpha\right), \tag{91}$$

$$b_{1/2} = \frac{1}{4}\left(3\beta_{21}^2 + 3\beta_{22}^2 + \beta_{11}^2\cos\alpha + \beta_{12}^2\cos\alpha\right), \tag{92}$$

$$c_{1/2} = -\frac{1}{2}\sin\alpha\,\beta_{11}\beta_{12}, \tag{93}$$

where $d\phi = \omega_1 dt$. Note that as A does not depend on time, the final solution does not require $\mathcal{T}$-ordering. It should be also emphasized here that in the non-adiabatic case we discuss the term $A_{3/2}$ contains non-diagonal terms that is not the case when the adiabaticity condition is held [58]. Now the solution of the problem takes a particular form of (45):

$$\psi(t) = e^{-i\phi_n(t)} \, e^{-i\omega_1 t A} \, \psi(0), \tag{94}$$

Formula (94) solves the problem of the evolution control for the system under consideration. The resulting quantum gate is entirely determined by A and the evolution law of the magnetic field, i.e. by a contour in the parameter space. Of course it is always possible to choose the parameters so that A turns out to generate a 2-qubit transformation that produces a superposition of basis states. For this reason the gate can be thought of as a universal one [14]. Of course, a suitable speed of the parameters evolution can not be reached by rotation of the sample as it took place in the experiment by authors of [71]. Nevertheless it is clear that this manner of control is not principle and one could imagine a situation where the parameters evolution is provided by the controlling magnetic field by adding a non-stationary transverse component. It should be also noted here that general formulas (85) do not provide an apparent way to realize CNOT gate or another common 2-qubit gate. They just give the evolution law of the system provided that the external parameters vary as shown above. To knowledge of the authors other examples of computation of a conditional geometric phase that would be both non-Abelian and non-adiabatic are absent. To provide the gates of common interest one has to invent some special case of the parameters variation which makes the generic evolution operator more simple and transparent. This subject is out of the scope of this article.

4. Conclusion

The approach developed in [75]– [77] is to be applied in the models where it is not possible to reduce the computation of the geometric phase to the case of 2-level system. Among those relevant to QC one can point out e.g. the model with anisotropic Heisenberg ferromagnetism where the exchange term in (51) is determined by a matrix of constants J_{ab} rather than by a single constant J. In this case the Zeeman terms $H_S^{(0)}$ no longer commute with the exchange term $J_{ab} S_a \otimes S_b$ and to derive the expression for the geometric phase it is necessary to consider a more general case of 4-level system. The problem becomes more complicated also if the superfine electron-nucleus spin interaction must be taken into account. It is the case for the Kane model of silicon QC [79]. The effective dimension of the system's Hamiltonian is then 16. It is hopeless to attempt to obtain an exact analytic expression for the system's dynamics which should be investigated numerically (see e.g. [78]) but it is nevertheless possible to derive an exact formula at least for the adiabatic phase.

One more problem to be mentioned here is interaction with the environment. It can appear to be important not only for such issue as decoherence but it also can in principle contribute to the geometric phase. The simplest way to see it is to consider the model described by (50)– (54). If the external electromagnetic wave field can affect not only the qubits but the phonons as well. The phonon degrees of freedom can produce Heisenberg–Weyl-like geometric phase that can fill the sign of the spin projection due to electron-phonon term (53), (54). Some more complicated interaction between the qubits and the environment makes it necessary to compute the geometric phase for a system with the symmetry algebra which is larger than $su(2)$ and cannot be reduced to the last one (in the sense of the phase derivation).

Other field of application could be multi-beam optical schemes for quantum computations where several energy levels must be included in the scheme to provide two-qubit operations. Besides of some special cases [80] it can require more general methods for the geometric phase computation.

References

[1] Deutsch, D. *Proc. Roy. Soc. London* 1985, A400, 97-117; Ibid. 1989, A425, 73-90

[2] Feynman, R. *Int. J. Theor. Phys.* 1982, 21, 467-488

[3] Steane, A.M. *Rep. Progr. Phys.* 1998, 61, 117-173; (1997) *Quantum Computing.* quant-ph/9708022

[4] Cabello, A. (2000) *Bibliographic guide to the foundations of quantum mechanics and quantum information.* quant-ph/0012089

[5] Kilin, S.Ya. *Progres in Optics* 2001, 42, 1-91

[6] Lloyd, S. *Science* 1993, 261, 1569-1571; Ibid. 1994, 263, 695-697

[7] Bermen, G.P.; Doolen, G.D.; Holm, D.D.; Tsifrinovich, V.I. *Phys. Lett.* 1994, A193, 444-450

[8] Barenco, A.; Deutsch, D.; Ekert, E.; Josza, R. *Phys. Rev. Lett.* 74 (1995) 4083-4086

[9] Di Vincenzo, D.P. *Science* 1995, 270, 255-257

[10] Turchette, Q.A.; Hood, C.J.; Lange, W.; Mabuchi, H.; Kimble, H.J. *Phys. Rev. Lett.* 1995, 75, 4710-4713

[11] Monroe, C.; Meekhof, D.M.; King, D.E.; Itano, W.M.; Wineland, D.J. *Phys. Rev. Lett.* 1995, 75, 4714-4717

[12] Cory, D.G.; Fahmy, A.F.; Havel, T.F. *Proc. Nat. Acad. Sci.* 1997, 94, 1634-1641; *Physica* 1998, D120, 82-91

[13] Lloyd, S. *Phys. Rev. Lett.* 1995, 75, 346-349

[14] Deutsch, D.; Barenco, A.; Ekert, A. *Proc. Roy. Soc. London* 195, A449, 669-667

[15] Kitaev, A.Yu. (1997) *Fault-tolerant quantum computation by anyons.* quant-ph/9707021

[16] Preskill, J. (1999) *Quantum information and physics: some future directions.* quant-ph/9904022

[17] Corac, J.I.; Zoller, P. *Phys. Rev. Lett.* 1995, 74, 4091-4094

[18] Gershenfeld, N.A.; Chuang, N.L. *Science* 1997, 275, 350-156

[19] Zanardi, P.; Rasetti, M. *Phys. Lett.* 1999, A264, 94; quant-ph/9904011;

[20] Pachos, J.; Zanardi, P.; Rasetti, M. *Phys. Rev.* 2000, A61, 1-8; quant-ph/9907103

[21] Pachos, J.; Chountasis, S. *Phys. Rev.* 2000, A 62, 2318-2324; quant-ph/9912093

[22] Ekert, A. Ericsson, M. and Hayden, P. (2000) *Geometric Quantum Computation.* quant-ph/0004015

[23] Fuentes-Guridi I., Bose S. and Vedral V. (2000) Proposal for measurment of harmonic oscillator Berry phase in ion traps. quant-ph/0006112; *Phys. Rev. Lett.* 2000, 85, 5018-5020

[24] Pellizzari, T.; Gardiner, S.A.; Cirac, J.I.; Zoller, P. *Phys. Rev. Lett.* 75 (1995) 3788-3791

[25] Averin, D.V. *Solid State Commun.* 1998, 105, 659-664

[26] Xiang-Bin, W.; Kieji, M. (2001) *Nonadiabatic conditional geometric phase shift with NMR.* quant-ph/0101038

[27] Xiang-Bin, W.; Kieji, M. (2001) *NMR C-NOT gate through Aharonov-Anandan's phase shift.* quant-ph/0105024

[28] Xiang-Bin, W.; Kieji, M. (2001) *On the nonadiabatic geometric quantum gates.* quant-ph/0108111

[29] Shi-Liang Zhu; Wang, Z.D. *Implementation of universal quantum gates based on nonadiabatic geometric phases.* quant-ph/0207037

[30] Du, J.; Shi, M.; Wu, J.; Zhou, X.; Han, R. (2002) *Implementation of nonadiabatic geometric quantum computation using NMR.* quant-ph/0207022

[31] Oshima, K; Azuma, K. (2003) *Proper magnetic fields for nonadiabatic quantum gates in NMR.* quant-ph/0305109

[32] Blais, A.; Tremblay, A.M.S. (2003) *Effect of noise on geometric logic gates for geometric quantum computation.* quant-ph/0105006

[33] Marinov, M.S.; Strahov, E. (2000) *A geometrical approach to non-adiabatic transitions in quantum theory: application to NMR, over-barrier reflection parametric excitayion of quantum oscillator.* quant-ph/0011121

[34] Zhang, X.D.; Zhu, S.L.; Hu, L.; Wang, Z.D. (2005) *Non-adiabatic geometric quantum computation using a single-loop scenario.* quant-ph/0502090

[35] Das, R.; Kumar, S.K.K.; Kumar, A. (2005) *Use of non-adiabatic geometric phase for quantum computing by nuclear magnetic resonance.* quant-ph/0503032

[36] Solinas, P.; Zanardi, P.; Zanghi, N.; Rossi, F. (2003) *Non-adiabatic geometrical quantum gates in semiconductor quantum dots.* quant-ph/0301089

[37] Solinas, P.; Zanardi, P.; Zanghi, N.; Rossi, F. (2003) *Holonomic quantum gates: a semiconductor-based implementation.* quant-ph/0301090

[38] Li, X.; Cen, L.; Huang, G.; Ma, L.; Yan, Y. *Non-adiabatic quantum computation with trapped ions.* quant-ph/0204028

[39] Scala, M.; Militello, B.; Messina, A. (2004) *Geometric phase accumulation-based effects in the quantum dynamics of an anisotropically traped ion.* quant-ph/0409168

[40] Zhu, S.L.; Wang, Z.D. (2002) *Geometric phase shift in quantum computation using superconducting nanocircuits: nonadiabatic effects.* quant-ph/0210175

[41] Nazir, A.; Spiller, T.P.; Munro, W.J. (2001) *Decoherence of geometric phase gates.* quant-ph/0110017

[42] Brion, E.; Harel, G.; Kebaili, N.; Akulin, V.M.; Dumer, I. (2002) *Decoherence correction by the Zeno effect and non-holonomic control.* quant-ph/0211003

[43] Carollo, A.; Fuentes-Guridi, I.; Santos, M.F.; Vedral, V. (2003) *Spin-1/2 geometric phase driven by quantum decohering field.* quant-ph/0306178

[44] Fuentes-Guridi, I; Girelli, F.; Livine, E. (2003) *Holonomic quantum computation in the presence of decoherence.* quant-ph/0311164

[45] Gaitan, F. (2003) *Noisy control, the adiabatic geometric phase and destruction of the efficiency of geometric quantum computation.* quant-ph/0312008

[46] Yi, X.X.; Wang, L.C.; Wang, W. (2005) *Geometric phase in dephasing systems.* quant-ph/0501085

[47] Pachos, J.; Zanardi, P. (2000) *Quantum holonomies for quantum computing.* quant-ph/0007110

[48] Lucarelli, D. (2002) *Control algebra for holonomic quantum computation with squizeed coherent states.* quant-ph/0202055

[49] Niskanen, A.O.; Nakahara, M.; Salomaa, M.M. (2002) *Realization of arbitrary gates in holonomic quantum computation.* quant-ph/0209015

[50] Tanimura, S.; Hayashi, D.; Nakahara, M. (2003) *Exact solutions of holonomic quantum computation.* quant-ph/0312079

[51] Nordling, M.; Sjöqvist, E. (2004) *Mixed-state non-Abelian holonomy for subsystems.* quant-ph/0404162

[52] Zhang, P.; Wang, Z.D.; Sun, J.D.; Sun, C.P. (2004) *Holonomic quantum computation using Rf-SQUIDs Coupled through a microwave cavity.* quant-ph/0407069

[53] Yi, X.X.; Chang, J.L. (2004) *Off-diagonal geometric phase in composite systems.* quant-ph/0407231

[54] Whitney, M.S.; Makhlin, Yu.; Shnirman, A.; Gefen, Y. (2004). *Geometric nature of the environment-induced Berry phase and geometric dephasing.* quant-ph/0405267

[55] Messiah, A. *Quantum Mechanics*; North Holland: Amsterdam, 1961; V.2

[56] Berry, M.V. *Proc Roy Soc London* 1984, A 392,35-47

[57] Wilczek, F.; Zee, A. *Phys. Rev. Lett.* 1984, 52, 2111-2114

[58] Moody, J.; Shapere, A.; Wilczek, F. *Phys. Rev. Lett.* 1984, 56, 893-895

[59] Shapere, A.; Wilczek F. (eds.) *Geometric Phases in Physics*; World Scientific: Singapore, 1989

[60] Aharonov, Y.; Anandan, J. *Phys. Rev. Lett.* 1987, 58, 1593-1595

[61] Anandan, J. Stodolsky, L. *Phys. Rev.* 1987, D35, 2597-2600

[62] Tolkachev, E.A.; Shnir Ya.M.; Tregubovich, A.Ya. In *Topological phases in Quantum Theory*; Dubovik, V.M.; Markovski, B.L.; Vinitski, S.I.; Ed.; World Scientific: Singapore, 1989; pp 119-128

[63] Tolkachev, E.A.; Boukanov, I.V.; Tregubovich A.Ya. *Phys. Atom. Nucl.* 1996, 59, 659-661

[64] Chaturvedi, S; Sriram, M.S.; Srinivasan, V. *J. Phys.* 1987, A20, L1071-L1075

[65] Chiao, R.Y.; Jordan, T.F. *Phys. Lett.* 1988, A132, 77-81

[66] Stoler, D. *Phys. Rev.* 1970, D1, 3217-3229

[67] M.Goto, F.Grosshans, *Semisimple Lie algebras*; Marcel Dekker,Inc.: New York and Basel, 1978

[68] Helgason, S. Differential Geometry and Symmetric Spaces; Academic Press: New York and London, 1962

[69] Korenblit, S.E.; Kuznetsov, V.E.; Naumov, V.A. In *Quantum systems: new trends and methods, Proceedings*; Barut, A.O. et al.Ed.; World Scientific: Singapore, 1995; pp 209-217

[70] Tycko, R. *Phys. Rev. Lett.* 1987, 58, 2281-2283

[71] Appelt, S.; Wäckerle, G.; Mehring, M. *Phys. Rev. Lett.* 1994, 72, 3921-3924

[72] Appelt, S.; Wäckerle, G.; Mehring, M. *Phys. Lett.* A204 (1995) 210-216

[73] Appelt, S.; Wäckerle, G.; Mehring, M. *Z. Phys. D*, 1995, 34-45

[74] Caldeira, A.O.; Legett, A.J. *Phys. Rev. Lett.* 1981, 46, 211-213

[75] Margolin, A.E.; Strazhev, V.I.; Tregubovich, A.Ya. *Phys. Lett.* 2002, A303, 331-334

[76] Margolin, A.E.; Strazhev, V.I.; Tregubovich, A.Ya. *Phys. Lett.* 2003, A312, 296-300

[77] Margolin, A.E.; Strazhev, V.I.; Tregubovich, A.Ya. *Optica and Specroscopia* 2003, 94, 789-791

[78] Wellard, C.; Hollenberg, L.C.L.; Pauli, H.C. (2001) *A non-adiabatic controlled NOT gate for the Kane solid state quantum computer.* quant-ph/0108103

[79] Kane, B.E. *Nature* 1998, 393, 133

[80] Cen, L.; Li, X.; Yan, Y. Zheng, H.; Wang, S. (2002) *Evaluating holonomic quantum computation: beyond adiabatic limitation.* quant-ph/0208120

In: High Energy Physics Research Advances
Editors: T.P. Harrison et al, pp. 137-167

ISBN: 978-1-60456-304-7
© 2008 Nova Science Publishers, Inc.

Chapter 6

ULTRA HIGH ENERGY COSMIC RAYS: IDENTIFICATION OF POSSIBLE SOURCES AND ENERGETIC SPECTRA

A.V. Uryson

Lebedev Physics Institute of Russian Academy of Sciences, Moscow 119991

Abstract

In this chapter origin of ultra high energy cosmic rays is discussed, namely possible sources, processes in which particles are accelerated, and energetic spectra. The chapter includes five sections. In Section 1 different hypotheses of ultra high energy cosmic rays origin are reported and the list of arrays for their detection is given. From our point of view, cosmic ray sources may be active galactic nuclei. Then it is possible to identify cosmic ray sources directly. The identification procedure is described and results of identification are presented in Section 2. It appears that potential sources are active galactic nuclei with red shifts z<0.01 and Blue Lacertae objects. If this is the case the question is how particles are accelerated to energies more than 10^{20} eV in these objects. This problem is discussed in Section 3. The process of charged particles acceleration in Blue Lacertae objects was suggested elsewhere. Here conditions in active galactic nuclei with moderate luminosities are discussed and possible mechanism of particle acceleration up to 10^{21} eV is considered. In extragalactic space particles interact with cosmic microwave background radiation and therefore unevitably loose energy. As a result a black body (GZK) cut off can appear in the cosmic ray spectrum at ultra high energies. Nearby active nuclei are located at distances less than 40 Mpc (if Hubble constant is 75 km/Mpc·s), and cosmic rays reach the Earth without significant energy losses. Blue Lacertae objects have red shifts from z=0.02 up to z>1 and it is unclear if particles accelerated in these objects can reach the Earth at energies $3 \cdot 10^{20}$ eV, that is the maximal energy detected in cosmic rays. Can our model explain the measured spectrum? Cosmic ray spectra at ultra high energies are considered in Section 4. Spectra of cosmic protons at the Earth are calculated and are compared with the measured one. In addition the limit on maximal cosmic ray energy is derived. In Section 5 main results are listed and predictions of different models are compared.

1. Introduction

This chapter deals with cosmic rays (CR) at ultra high energies (UHE), $E>4\cdot10^{19}$ eV. UHECRs are generally believed to be extragalactic in origin (see e.g. Hayashida et al. 1996; Hillas, 1998), but their sources have not been firmly established. Sources considered in the literature can be divided into three categories. The first one includes astrophysical objects, such as pulsars, active galactic nuclei (AGN), the hot spots and cocoons (these are components of jets) of powerful radio galaxies and quasars, gamma-ray bursts, and interacting galaxies (Berezinsky et al. 1990; Bhattacharjee and Sigl, 2000; Nagano and Watson, 2000; references therein). The second category of proposed UHECR sources is cosmic topological defects (Hill et al. 1987; Berezinsky and Vilenkin, 1997), and the third is the decay of supermassive metastable particles of cold dark matter that have accumulated in galactic halos (Kuzmin and Rubakov, 1998; Berezinsky et al., 1997). Direct identification of astrophysical objects is possible only in the first case. In the second case, any objects falling within error boxes centered on particle arrival directions should be chance coincidences. In the third case, most of the particle flux should arrive from the Galactic halo and, perhaps, partly from the Virgo cluster of galaxies (Berezinsky and Vilenkin, 1997).

In this chapter the hypothesis of UHECR origin in AGNs is considered. This idea is not new, it was widely discussed in the 1970s. Some estimates of AGN power and of their spatial density were derived which illustrated the validity of the idea (Berezinsky et al., 1990).

If UHECR sources are astrophysical objects it may be possible to identify them. The main assumption underlying the identification is that UHECRs fly rectilinearly from sources to the Earth. What are difficulties of the identification procedure? The CR intensity exponentially decreases with increasing energy, and CR flux at energies $E\approx10^{20}$ eV is only 1 particle/(100 km^2 year). So it is a problem how to gather statistics of UHECRs required for identification. How UHECRs are detected? CR particles initiate external air showers in the atmosphere and because of this arrays which detect showers are used for UHECR registration and analysis. Giant arrays, at which CR detectors are located in the regions of tens and hundreds square kilometers were created. These arrays are the following (Berezinsky et al. 1990; Nagano and Watson, 2000):

Volcano Ranch (USA, operated in 1959-1963), 8 km^2;
Haverah Park (England, operated from 1967), 12 km^2;
SUGAR (Sydney University Giant Array, Australia, 1974-1982), 60 km^2;
Yakutsk array (operates since 1974), 18 km^2, and 10 km^2 since 1996;
Akeno (Japan, operates since 1979), 1 km^2, and AGASA (Akeno Giant Shower Array, Japan), 100 km^2;
Fly's Eye (USA, operated in 1981-1992) with two stations at distance 3.4 km.
HiRes (USA, High Resolution) has two stations at distance 12.5 km;
Pierre Auger observatory includes two arrays in North and South hemispheres each having square of 3000 km^2.

In addition UHECR detecting onboard satellites is suggested. In such measurments radio signal that is produced by external shower in the atmosphere will be detected and large sections of the atmosphere will be observed (Chechin et al., 2002).

The axis of a shower indicates the arrival direction of the primary CR particle (showers, in which this is not realized seems to be extremely rare). Errors in UHECR arrival directions are of 3-10^0 (AGASA has errors of 1.5-3^0), the typical energetic resolution is $\Delta E/E \approx 0.3$. In Auger observatory the angular resolution is of 1^0, and the energetic resolution is of 0.1.

Errors in particle arrival directions make it difficult to identify UHECR sources. Another difficulty in identification procedure is UHECR deflection in magnetic fields on their way to the Earth. (Here we do not discuss completness of catalogues of AGNs, that also effects on identification.)

In spite of difficulties encountered in identification procedure we performed identification in our paper (Uryson, 1996). We analysed showers initiated by UHECRs which arrival directions along with their errors were published. In that paper the statistics of UHECRs considered was 17. It appeared that sources were nearby Seyfert nuclei having red shifts z≤0.0092 (i.e. located at distances less than 40 Mpc if Hubble constant is H=75 km/(s Mpc)). Later we found (Uryson, 2001a,b) that Blue Lacertae objects (BL Lac's) are possible UHECR sources too. BL Lac's were identified as possible sources also by Tinyakov and Tkachev (2001). Since the paper (Uryson, 1996) we repeated the identification procedure with larger statistics of UHECRs and with refreshed catalogues of AGNs.

The next step of our investigation was to explain how particles are accelerated in the sources to energies E>4·10^{19} eV. Possible mechanism of CR acceleration to extremely high energies in AGNs was suggested by Kardashev (1995). His paper deals with AGNs having powerful jets which are likely to be BL Lac's. However Seyferts identified as posssible UHECR sources have moderate luminosities and weak jets. We analysed conditions in such AGNs and proposed a model for the CR acceleration up to 10^{21} eV. (The highest energy registered in CR is 3·10^{20} eV (Bird et al., 1995).)

The next problem was to analyse UHECR propagation in the extragalactic space.

The space is filled with cosmic microwave background radiation, with which CRs interact losing energy. The reactions with photons are threshold. Energy losses increase abruptly at the energy about 4·10^{19} eV and in consequence the number of particle at E>4·10^{19} eV decreases. As a result the CR spectrum at UHE may have a cut off. This was analysed by Greizen (1966) and by Zatsepin with Kuz'min (1966), and it is called black body or GZK cut off. The UHECR data do not show GZK cutoff except HiRes data (Bergman, 2007). However, the spectrum has a rather complex shape at energies 10^{17}-10^{20} eV, which could be interpreted as a cutoff (Bahcall and Waxman, 2003).

GZK cutoff should not be observed if sources of UHECRs are relatively nearby objects: the mean free path of particles with energies $E<10^{20}$ eV in the background radiation field is about ~40-50 Mpc, and particles with energies up to $E \approx 10^{21}$ eV should traverse distances of about 10-15 Mpc essentially unattenuated (Stecker 1968; 1998).

Nearby AGNs are located at distances less than 40 Mpc, and CRs reach the Earth without significant energy losses. BL Lac's are remoted at distances to 1000 Mpc, and it is unclear if particles accelerated in BL Lac's can reach the Earth at the highest energy 3·10^{20} eV. We computed spectra at the Earth of cosmic protons that were accelerated in Seyferts and in BL Lac's and compared them with the measured spectrum. It appeared that our model described the measured spectrum at E>4·10^{19} eV.

The structure of the chapter is the following. In Section 2 we present the identification procedure and results of identification. UHECR deflection in magnetic fields on the way from sources to the Earth is analysed and conditions under which the identification procedure is correct are discussed. In Section 3 data for AGNs is considered. It is shown that the conditions required for the acceleration and for free escape of particles exist in Seyfert galaxies. The model for UHECR acceleration in AGNs with weak jets is described and some other models are discussed. In this Section we derive also estimates of the energy of Seyferts based on the UHECR spectrum observed at the Earth. In Section 4 we consider spectra of cosmic protons accelerated in Seyferts and in BL Lac's. We present estimates of the energy of BL Lac's required for CR acceleration. In the Section 5, we list results obtained and consider the characteristics of CRs predicted by different models.

2. Identification of Nearby Active Galaxies as Sources of CR above 4×10^{19} EV

2.1. Introduction

Different models of UHECR origin are discussed in literature, as it is refered in Section 1. Here we adopted the model that CR sources are astrophysical objects as our working hypothesis. We supposed that possible sources could be X-ray pulsars (i.e., the most powerful ones), Seyfert galaxies, BL Lac's, or radio galaxies. We further assumed that particle trajectories in intergalactic space are almost rectilinear and ignored particle deflection by magnetic fields in the Galaxy. We searched for possible sources within error boxes centered on the arrival directions of showers, and calculated the probabilities of objects being in the error boxes by chance.

In identification procedure we consider Seyferts at red shifts $z<0.01$. This choice is resulted from our paper (Uryson, 1996). As a rule, more than one Seyfert was present in the error range of arrival direction of a shower, but within each range there was at least one galaxy with red shift $z\leq0.0092$. The calculated probability P of the random occurrence of such Seyferts in the search field of a given number of showers was less than 10^{-4}, so random occurences were unlikely events. Nearby Seyferts being UHECR sources, it explains that more than ten UHECRs at $E\geq10^{20}$ eV were detected at different arrays. We used catalogues of Seyferts by Veron-Cetty and Veron (1991) and by Lipovetsky et al (1987) in this paper.

For pulsars and radio galaxies the probability of random occurence P was about 0.1. In (Uryson 2001a,b) we found that P is also small for BL Lac's. We did not get this result previously using the catalogue by Veron (1993), where the statistics of BL Lac's at $\delta>-10^{0}$ was 55. We identified BL Lac's with the catalogue by Veron (1998), where the statistics of BL Lac's tripled, 159. The result that Bl Lac's are probable sources of UHECRs were also got by Tinyakov & Tkachev (2001) and by Gorbunov et al. (2002).

In none of our papers have we supposed the possible sources to be interacting galaxies, where conditions can exist for the CR acceleration (Cesarsky and Ptuskin, 1993). The reason is that interacting galaxies are the majority of normal galaxies (Wright et al., 1988), and their number is tens of times greater than the number of AGNs. Consequently, the probability of random incidence of any normal galaxy in the search field will be high. At the present time it is difficult to select from the observational data galaxies in which conditions exist for CR acceleration.

Other results on UHECR sources identification are the following. Farrar and Biermann (1998) found quasars to be UHECR sources. However this result was discussed by Hoffman (1999) and by Sigl et al. (2001). Sigl et al. (2001) showed that neither compact radio sources nor gamma-ray emitting blazars seemed to be sources of UHECRs.

Here we present identification of possible UHECR sources with statistics of 63 showers. We used the catalogue (Veron-Cetty and Veron 2001) to search for AGNs. We consider also UHECR deviation in magnetic fields on their way to the Earth.

2.2. The Identification Procedure

We used showers whose arrival directions were published along with their errors. For identification procedure, we selected showers with errors in arrival directions ($\Delta\alpha$, $\Delta\delta)\leq3^0$ in equatorial coordinates: 58 events with $E>4 \cdot 10^{19}$ eV (Takeda et al., 1999; Hayashida et al., 2000), 4 Yakutsk showers with $E>4\cdot10^{19}$ eV (Afanasiev et al., 1996), and 1 Haverah Park shower with $E\geq10^{20}$ eV (Watson 1995), its error was computed in (Farrar and Biermann 1998). Of these showers, 11 have energies $E>10^{20}$ eV. The most energetic shower with the energy $E\approx3\cdot10^{20}$ eV (Bird et al. 1995) was not included in the sample considered here because of the large coordinate errors: $\alpha=85.2\pm0.4^0$, $\delta=48.0(+5.2, -6.3)^0$.

Different objects occur around the patricle arrivals. We use the following procedure to obtain the probability of a chance occurence of objects near shower arrivals. The showers were subdivided into several groups depending on Galactic latitude b of arrival directions, and in each error box we looked for objects of the given type. We counted the number of showers K in each group and the number of showers N which have at least one object of the given type within the error box. The showers were subdivided in Galactic latitude b in order to exclude events clearly lying in the galaxy 'avoidance zone'. (These zones are low-latitude regions of sky, where galactic gas and dust make it difficult to observe objects. By this reason catalogues contain less objects at low b in comparison with $|b|\geq20\text{-}30^0$.) We calculated the probabilities that objects of the given type would fall in the fields of search of N of the total of K showers by chance as follows. Showers with randomly distributed arrival coordinates were simulated. The coordinates of the simulated showers were determined by a random-number generator (Forsythe et al., 1977) within a survey band $\alpha=$ 0-24 hr and $\delta=$ -10-90°. (This band is close to bands of abovementioned arrays.) We subdivided simulated showers into groups in the same way as real showers. Each simulated group contains the number K of showers equal to those observed. We then counted in each simulated group the number of showers N_{sim} having at least one object of the given type located in the error box (N_{sim} can take values in the interval $0\leq N_{sim}\leq K$). Next we counted the number of simulated groups I_{sim} having a given number of showers N_{sim}. In a group of K showers, the probability P of a chance occurrence of galaxies in the field of search of a given number of showers N_{sim} was determined as

$$P = \sum_{i=1}^{M}(I_{sim})_i / M$$

where $M=10^5$ is the number of trials performed for each group. By the law of probability the coincidence is by chance if the probability P is $P<3\sigma$, where σ is the parameter of Gaussian distribution of P values.

What is the size of the region of search? Statistics and the law of probability give the following data (Hudson, 1964; Squires, 1968): the probability that the particle coordinates are within the one-mean-square error box is 68 per cent, the probability for the two-mean-square error box is 95 per cent, and for the three-mean-square error box it is 99.8 per cent. This means that more than 30 per cent of the objects are excluded from the analysis a priori using 1-error box, only 5 per cent of objects are lost with 2-error box, and essentially all objects are considered with 3-error box. Using 2-error box is less strict than using 3-error box, and it is more accurate as compared with 1-error box. One believes that using 1-error box region would reduce the chance coincidence against 2- and 3-error boxes.

In the papers by Farrar and Biermann (1998), by Sigl et al. (2001), by Tinyakov and Tkachev (2001), and by Gorbunov et al. (2002) 1-error box was used. We here use both 1-, 2-, and 3-error boxes in the identification procedure.

The optical coordinates of the galaxies and pulsars are accurate to several arcseconds, so the fields of search were determined solely by the errors in the shower coordinates.

2.3. Results

2.3.1. Nearby Active Galaxies

The catalogue (Veron-Cetty and Veron, 2001) contains both Seyferts with detailed classification and objects which are probably or possibly Seyferts, because of a lack of avaliable data. We found the probabilities of chance coincidence in two cases: for all Seyferts and for Seyferts with detailed classification having $z<0.01$.

For all nearby Seyferts the probabilities of chance coincidence using 1-, 2-, and 3-error boxes $P_1(N)$, $P_2(N)$, and $P_3(N)$ are the following (N is the number of showers having at least one nearby galaxy within the error box; if some showers in a group had galaxies in regions smaller than 3-error box but larger than 2-error box, we determined the weighted mean error box; for example 2-error box means also 2.1- or 2.2-error box, and similarly 1-error box means also 1.2- or 1.3-error box):

63 showers with no selection in Galactic latitude, P_1 (16)$=1.1 \cdot 10^{-3}$, P_2 (27)$=3.6 \cdot 10^{-4}$, P_3(29)$=2.4 \cdot 10^{-2}$;

54 showers with $|b|>11.2^0$ P_1(16)$=1.2 \cdot 10^{-3}$, P_2(26)$=6.5 \cdot 10^{-4}$, P_3(29)$=1.8 \cdot 10^{-2}$;

37 showers with $|b|>21.9^0$ P_1(13)$=3.2 \cdot 10^{-3}$, P_2(23)$=1.8 \cdot 10^{-4}$, P_3(23)$=2.5 \cdot 10^{-2}$;

27 showers with $|b|>31.7^0 P_1$(14)$=5.1 \cdot 10^{-4}$, P_2(23)$=2.0 \cdot 10^{-5}$, P_3(23)$=9.5 \cdot 10^{-3}$.

Here probabilities are $P>3\sigma$ for 1- and 2- error boxes for showers at any latitudes, except $|b|>21.9^0$, where $P_1 \approx 2.95\sigma$.

For nearby Seyferts having detailed classification, the probabilities are:

63 showers with no selection in Galactic latitude, P_1(12)$=1.1 \cdot 10^{-2}$, P_2 (23)$=3.2 \cdot 10^{-3}$, P_3 (27)$=3.2 \cdot 10^{-2}$;

54 showers with $|b|>11.2^0$ P_1(12)$=1.5 \cdot 10^{-2}$, P_2(22)$=5.2 \cdot 10^{-3}$, P_3 (27)$=2.3 \cdot 10^{-2}$;

37 showers with $|b|>21.9^0$ P_1(9)$=3.0 \cdot 10^{-2}$, P_2(19)$=3.0 \cdot 10^{-3}$, P_3 (21)$=3.7 \cdot 10^{-2}$;

27 showers with $|b|>31.7^0 P_1(10)=1.0\cdot10^{-2}$, $P_2(19)=1.1\cdot10^{-3}$, $P_3(21)=2.2\cdot10^{-2}$.

Here probabilities are $P \geq 3\sigma$ for 2-error boxes at any latitudes, except showers at $|b|>11.2^0$ where $P_2 \approx 2.80\sigma$.

Because of low values of probabilities it is difficult to ignore nearby AGNs as possible UHECR sources. (Probabilities increase with increasing z, see (Uryson, 1996) for $P(z)$ relations for showers arriving from sky areas located at arbitrary b.)

2.3.2. Blue Lacertae Objects

Confirmed, probable or possible BL Lac's are listed in (Veron-Cetty and Veron, 2001). For all these objects the probabilities of chance coincidence using 1-, 2-, and 3-error boxes $P_1(N)$, $P_2(N)$, and $P_3(N)$ are the following.

63 showers with no selection in Galactic latitude, $P_1(45)<4\cdot10^{-5}$, $P_2(56)=6.\cdot10^{-5}$, $P_3(57)=2.3\cdot10^{-2}$;
54 showers with $|b|>11.2^0 P_1(42)<3.\cdot10^{-5}$, $P_2(51)=3.4\cdot10^{-4}$, $P_3(51)=1.0\cdot10^{-1}$;
37 showers with $|b|>21.9^0 P_1(36)<4.\cdot10^{-5}$, $P_2(36)=2.5\cdot10^{-3}$, $P_3(36)=8.7\cdot10^{-2}$;
27 showers with $|b|>31.7^0 P_1(27)<4.\cdot10^{-5}$, $P_2(27)=5.2\cdot10^{-2}$, $P_3(27)=4.5\cdot10^{-1}$.

For confirmed BL Lac objects, the probabilities are:

63 showers with no selection in Galactic latitude, $P_1(38)<1.\cdot10^{-5}$, $P_2(48)=4.7\cdot10^{-4}$, $P_3(53)=1.0\cdot10^{-2}$;
54 showers with $|b|>11.2^0 P_1(42)<3.\cdot10^{-5}$, $P_2(43)=6.9\cdot10^{-3}$, $P_3(47)=7.5\cdot10^{-2}$;
37 showers with $|b|>21.9^0 P_1(28)<3.\cdot10^{-5}$, $P_2(33)=1.8\cdot10^{-3}$, $P_3(35)=4.0\cdot10^{-2}$;
27 showers with $|b|>31.7^0 P_1(24)<1.\cdot10^{-5}$, $P_2(26)=1.2\cdot10^{-2}$, $P_3(27)=2.0\cdot10^{-1}$.

BL Lac's are considered as possible UHECR sources due to low values of probabilities for 1-error boxes and in some cases for 2-error boxes. Similar results are obtained by Tinyakov and Tkachev (2001), and by Gorbunov et al. (2002).

2.3.3. X-ray Pulsars and Radiogalaxies

We searched for X-ray pulsars using the catalogue (Popov, 2000). The probability of the pulsars in the shower error boxes falling there by chance is ~0.01-0.1, consistent with these pulsars being chance coincidences.

Powerful radio galaxies are considered to be possible CR sources due to their high power. In some models (Norman et al., 1995), particles in radio galaxies can be accelerated to energies of $5\cdot10^{19}$–10^{20} eV. However, radio galaxies are also considered to be the most likely sources for particles with energies $2\cdot10^{20}$ and $3\cdot10^{20}$ eV. Stanev et al. (1995) and Stecker (1998) report attempts to identify these galaxies under various assumptions about the intensity of the intergalactic magnetic field and the atomic numbers and propagation distances of the particles. Rachen et al. (1993) also analyzed the chemical composition and spectrum of detected particles assuming that their sources were radio galaxies. We searched for radio galaxies using the catalogues (Kuhr et al. 1981; Spinrad et al. 1985). Our identification list included all galaxies from (Spinrad et al. 1985) and galaxies with $z \leq 0.1$ from (Kuhr et al. 1981) that were absent from the first one. The probabilities of the radio galaxies with any z and $z < 0.1$ falling in the error boxes of shower by chance are $P \approx 0.01-0.1$, for showers arriving from fields of sky with various b values. Thus, the radio galaxies in the fields of search could represent chance coincidences.

2.4. Particles Deviation in Magnetic Fields

When identifying UHECR sources, we assumed that the particles propagate almost rectilinearly through intergalactic space. The searches for possible sources were carried out in circles of radius up to $(3\Delta\alpha, 3\Delta\delta)=9^0$ centered on the shower axes. This is equivalent to assuming that the maximum deflection angle in the intergalactic magnetic field is $\alpha_o= 9°$.

At present we know neither the magnitude nor the character of inhomogeneities in the intergalactic magnetic field. Only theoretical upper limits for the magnetic-field intensity are available (Kronberg, 1994): $H<10^{-9}$ G based on measurements of the rotation measures of quasars with $z = 2.5$; $H<<10^{-9}$ G, if UHECR protons propagate rectilinearly through intergalactic space; and $H<10^{-11}$ G for the regular component of the magnetic field. Clusters of galaxies can harbor much stronger fields: $H\sim 10^{-6}-10^{-7}$ G at distances of up to ~0.5 Mpc from the cluster center.

We estimated the intensity of the intergalactic magnetic field outside clusters of galaxies for the simple case when particles move in the plane perpendicular to the magnetic field $\mathbf{H}$, assuming that $\alpha \approx 9^0$ and that field inhomogeneities can be neglected throughout the propagation path.

We looked for sources located in a given cone around the shower axis (Figure 1). The half-angle α of the cone is related to the arc L^* along which die particle moved from the source to the detection facility: $\alpha = L^*/2$. The length of the arc is $L= 2\alpha r_L$, where the Larmor radius is (Berezinsky et al., 1990):

$$r_B = E/(300ZH) \qquad (1)$$

Here, r_B is in cm, H is in G, and Z is the charge of the particle. Since $\alpha=\alpha_0$, we have $L/(2r_B)=\alpha_0$ and $H=(2\alpha_0E)/(300ZR)$. For angles $\alpha \leq 10^0$ one has $L \approx R$ where R is the distance between the source and the detector array. It follows that $H=(2\alpha_0E)/(300ZL)$, $\alpha_0=9^0$. We found above that most UHECR sources are located at distances $R<40$ Mpc. Therefore, we have $H \leq 8.7 \cdot 10^{-10}$ G for protons ($Z=1$) with energies $E \approx 10^{20}$ eV. If a fragment of heavy nuclei with such energies arrive from distances of 70 Mpc (Stecker, 1998), we have $H \leq 7 \cdot 10^{-11}$ G for $Z = 10$. These limits for the magnetic-field intensity H are consistent with the estimates in (Kronbeg, 1994).

We now consider the case when a UHECR source is a member of a cluster. According to data avaliable (private communication by Zasov), the sizes of galaxy clusters range from $D \approx 1$ Mpc to $D \approx 5-7$ Mpc. If the distance to the cluster is 40 Mpc, its angular size is $\alpha_{cl} \approx 1.4^0$ for $D \approx 1$ Mpc and $\alpha_{cl} \approx 10^0$ for $D \approx 7$ Mpc. In both cases, the cluster containing the source should fall in the 3-error box, since the angular size of the box exceeds that of the cluster, $2\alpha_0 \approx 18^0 > \alpha_{cl}$.

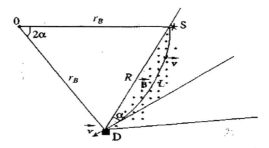

Figure 1. Motion of a particle in the plane perpendicular to the intergatactic magnetic field **H**. S is the source, D is the detector, O is the centre of the Larmor orbit, L is the arc along which the particle moves, and α is the angle between the direction toward the source and the shower axis (the direction of particle motion during detection).

If this estimate of extragalactic magnetic field strength is correct, we should search for sources within $(3\Delta\alpha, 3\Delta\delta) \approx 9^0$ areas also for showers with errors in arrival directions smaller than 3^0, $(\Delta\alpha, \Delta\delta) < 3^0$.

Now we treat UHECR deviation in the galactic magnetic field.

In the Galaxy, particles move in fields of $H \sim 10^{-6}$ G. In the disk, the magnetic field is regular in spiral arms and it is directed along them, and in the halo, the regular field component is perpendicular to the disk. The nonregular field component fluctuates with the main scale $\delta L \sim 100$ pc, $\delta H \sim 10^{-6}$ G (Kronberg, 1994). Various models are used to describe the large-scale galactic field structure (Cronin, 1996; Stanev, 1997; Bhattacharjee and Sigl, 2000). Following them deviations of charge particles in the ordered field depend on arrival directions and can be negligible. We estimate the rms angle of a particle deflection on path d in a random field from the relation (Cronin, 1996):

$$\psi \approx 1.7^0 \, (d/30 \text{ Mpc})^{1/2} (\delta L/1 \text{ Mpc})^{1/2} (\delta H/10^{-9}\text{G})(E/10^{20}\text{eV})^{-1}Z. \qquad (2)$$

Assuming the main scale of the halo magnetic field of $d \approx 2$ kpc one get

$$\psi \approx 0.014^0 (E/10^{20}\text{eV})^{-1}Z. \qquad (3)$$

It follows from (3) that particles with $Z \leq 26$ and $E \approx 10^{21}$ eV deviate at angles of $\psi < 0.4^0$, particles with $Z \leq 10$ and $E \approx 10^{20}$ eV deviate at angles of $\psi < 1^0$. These deflections have no effect on identification and may be disregarded.

Thus results presented are valid if, first, extragalactic magnetic fields outside galactic clusters do not exceed $H < 10^{-9}$ G, second, if particles arrived from directions where deviations in galactic regular fields are small, and third, if particles at energies $E \leq 10^{20}$ eV have charges $Z < 10$. (However, Stanev (1997) suggests that identification of possible UHECR sources is not meaningless even for large particle deflections in the galactic magnetic field: if the sources are actually astrophysical objects, then particles with $E > 10^{20}$ eV arrive from them with small deflections, while the particles with $E > 4 \, 10^{19}$ eV form a halo around their arrival directions;

so for large shower statistics it is possible to investigate the large-scale structure of the galactic magnetic field by analysing shower arrival directions and angles between them.)

3. Seyfert Nuclei as Sources of Ultrahigh-Energy Cosmic Rays

3.1. Some Models for CR Acceleration

Specific models for the acceleration of CR in AGNs were proposed by Haswell et al. (1992) and by Kardashev (1995). In the model of (Haswell et al., 1992), the magnetic field in an accretion disk surrounding a supermassive black hole evolves as a consequence of differential rotation of plasma with frozen-in magnetic field. As a result, there is an explosive (more rapid than exponential) growth in the electric field in some sections of the disk. Particles can be ejected from the flow of plasma in regions of low plasma density at the surface of the disk and can be accelerated by this strong electric field. The acceleration continues until the particles leave the region of explosive electric-field growth or until the growth in the field begins to slow. Numerical computation of the trajectories of individual particles show that particles in a disk with particle density $\sim 10^{16}$ cm^{-3} around a black hole with mass $\sim 10^7 M_O$ (M_O is the solar mass) can be accelerated to energies of $\sim 10^{21}$ eV. The accelerated particles lose some of their energy to synchrotron radiation and to the production of pairs in reactions with photons. Synchrotron losses will be negligible if the particle velocities are aligned with the magnetic field. (Acceleration in harmonic fields is possible in the case of other orientations of the particle velocities: if a particle is trapped between the crests of a wave and the magnetic field has a component perpendicular to the particle velocity, the particle will be accelerated (Sagdeev and Shapiro, 1973; Katsouleas and Dawson, 1983.) Losses to pair creation depend on the density distribution of the photons. In addition, in the case of collisions between particles and photons, these losses depend on the angle between the directions of motion of the particle and photon. It is therefore possible that some of the accelerated particles could leave the source galaxy virtually without any loss in their energy.

In the model of (Kardashev, 1995), the particle acceleration occurs in the electric field induced near a supermassive black hole with mass $\sim 10^9 M_O$ in periods of low activity of the black hole, when the accretion is diminished. The model is based on two main assumptions. The first is that the magnetic field of the black hole can be as high as $\sim 2 \cdot 10^{10}$ G, in contrast to the value $\sim 10^4$ G derived by Field and Rogers (1993), and the limit derived by Krolik (1999), $B^2 \sim 8\pi\rho$, where ρ is the density of matter in the accretion disk. Second, it is assumed that in quiescent states of the black-hole activity, there is a region with a very low plasma density, in which a very strong induced electric field cannot be compensated by the volume charge of the plasma. Regions in which such very strong electric fields can exist are located near the magnetic poles and rotational axis of the black hole. Particles can be accelerated to energies $10^{27} Z$ eV in these fields. The accelerated particles lose energy in interactions with photons in the disk and via curvature radiation. In quiescent phases, losses to direct pair production and losses in photon reactions are negligible. The particle energy can be decreased to $10^{21} Z$ eV due to curvature radiation, and the maximum proton energy is accordingly 10^{21} eV.

In both models, the particle acceleration occurs in the presence of extremely strong electric fields. The maximum particle energies $\sim 10^{21}$ eV coincide with the maximum energies detected for CRs (Bird et al., 1995).

The model of (Haswell et al., 1992) predicts sporadic flares of radiation associated with the ejection of accelerated particles, while the model of (Kardashev, 1995) predicts a flux of accelerated particles from the galactic nucleus. The Seyferts that have been identified as CR possible sources emit low radio and X-ray fluxes and do not display the characteristic features predicted by these models. In addition, it is not known whether there exist in these AGNs the conditions required to accelerate particles in the suggested scenarios. We suppose that various conditions can exist in the CR sources, with various acceleration mechanisms being realized. That is why we propose another model for the CR acceleration. Our model does not predict any specific features of the source radiation apart from the acceleration of particles to energies of $E \approx 10^{21}$ eV.

In CR sources, particles are not only accelerated to ultrahigh energies, but also leave the acceleration region without appreciable energy losses. We here show that the conditions required for free escape of particles exist in Seyferts with moderate luminosities. This section presents also estimates of the energy of Seyferts based on the UHECR spectrum observed at the Earth.

3.2. Conditions in Active Galactic Nuclei

The activity of AGNs is most likely due to the existence of a supermassive black hole with a mass of $M \approx 10^9 M_O$, and an associated accretion disk at the center of the galaxy. The accreting material is provided by gas from stellar winds and supernova explosions surrounding the nucleus, stellar remnants that are disrupted by tidal forces near the black hole, and entire stars that are captured by the hole.The thickness of the accretion disk appears to be ~1 pc (Antonucci, 1993). Models in which the disk thickness is ~10^{16} cm have also been discussed in the literature (Bednarek, 1993). The accretion disks in AGNs can apparently be either optically thin or optically thick (Antonucci, 1993; Begelman et al., 1984).

Two jets of material flowing out from the inner accretion disk emerge along the rotational axis. Jets of both ordinary and electron–positron plasma and fluxes of electromagnetic radiation have all been considered in the literature; the jet material can be ejected either in the form of individual "blobs" of plasma or as a continuous flow (Begelman et al., 1984). The material ejected from the inner regions of the disk flows through the disk along two funnel-like channels formed along the rotational axis.

The ejected particles interact with material in the funnel walls and with their radiation. As a result of pp and $p\gamma$ interactions, electron–positron pairs or pions are born, which decay into positrons and photons, so that an electromagnetic cascade is generated inside the torus, giving rise to collimated beams of gamma radiation (Bednarek, 1993; Begelman et al., 1984; Mannheim and Biermann, 1992; Sikora et al.,1994), providing an explanation for the gamma-ray emission of AGNs.

The conditions under which plasma from the disk and funnel walls does not fall into the channel are considered in (Bednarek and Kirk, 1995): this is hindered by a magnetic field of ~10^4 G oriented parallel to the rotational axis of the black hole. (As a result, the electric field inside the funnel cannot be compensated by the volume charge of the plasma, analogous to the model of (Kardashev, 1995), and particles are accelerated inside the funnel by this field.)

The central region of the galaxy is surrounded by a geometrically and optically thick dusty torus that radiates in the infrared; the thickness of the torus can reach ~100 pc (Pier and

Krolik, 1993). Models in which the inner walls of the torus are constantly bombarded by radiation, leading to the liberation of torus material that then falls into the jet, are considered in the literature (see (Falcke et al., 1995) and references therein).

As is clear from this brief summary of various models, disk material can either be contained in the jet from the beginning or fall into the jet at some point. We will assume that the jets contain plasma from the accretion disk.

Different types of AGNs differ in various ways, including the power of their radiation in various bands, the ratio of the radio luminosity to the luminosity of the accretion disk (radio "loudness" or radio "quietness"), and the strength of the jets. Radio images show that radio galaxies have jets extending to tens or hundreds of kiloparsecs (Fanaroff and Riley, 1974; Bridle and Perley, 1984), while the lengths of the jets in Seyfert nuclei are ~1–10 pc (Xu et al., 1999; Thean et al., 2000; Falcke et al., 2000; Nagar et al., 2001; Ulvestad and Ho, 2002). Seyfert galaxies are usually considered to be radioquiet objects (although it is possible that this is not the case for all types of Seyferts (Ho and Peng, 2001)).

How can we explain the moderate jets and radioquiet nature of Seyfert galaxies? Unified schemes for AGNs are discussed in the literature, in which observed differences in properties are due, for example, to differences in the accretion conditions (the mass of accreted gas and the accretion rate; (Rees, 1984)) the orientation of the AGN relative to the observer (Antonucci, 1993), the mass of the central black hole (Ghisellini and Celotti, 2001), etc. In addition, various authors have explained the radioloud/radioquiet dichotomy in various ways. It is possible that radio loudness is correlated with the spin of the central black hole: in radio-loud objects, the spin is close to its maximum value, while the spin is low in radio-quiet objects (Moderski et al., 1998). Falcke et al.(1995) propose a unified model for AGNs in which the different properties of radio-quiet and radio-loud objects, and also of FRI and FRII galaxies, are associated with different geometries for a torus surrounding the central black hole: the opening angle of the funnel is appreciably broader for radio-quiet objects and Seyferts than it is for radio-loud objects. As a consequence, the jets in radio-quiet objects and Seyfert galaxies are more poorly collimated and weaker and interact less with the torus (therefore, the production of electron– positron pairs in the funnel is small, so that these nuclei emit weakly in gamma rays). It is proposed in papers (Vilkoviskij and Karpova, 1996; Vilkoviskij et al., 1999) that, in the vicinity of the central black holes of radio-quiet objects, there are regions with fairly large stellar masses and colliding hot gas flows, in which the jets are disrupted (by ~90%) within ~1–3 pc of the nucleus.

A magnetic field is frozen in jet material. In some jets the field is oriented along them, and there are jets in which the field is directed across the jet (Gabuzda and Cawthorne, 2003). Magnetic fields in jets were investigated theoretically in (Istomin and Pariev, 1994).

Our model supposes that Seyferts have relativistic jets of ~1-3 pc in length, with magnetic field oriented along jets.

A cocoon of perturbed jet material is formed around the jet as it propagates (Begelman and Cioffi, 1989). Shock waves are excited in the jet and cocoon as a consequence of the nonlinear development of instabilities at the jet surface, collisions with dense clouds, and velocity fluctuations (Begelman et al., 1984; Blandford and Eichler, 1987; Chakrabarti, 1988). Relativistic particles can be accelerated to ultrahigh energies in the fronts of shocks with regular magnetic field (Krymsky, 1977; see the reviews (Hillas, 1984; Cesarsky, 1992) and references therein).

We suppose that CRs are accelerated in Seyfert nuclei in the fronts of shock waves in the jet. The magnetic-field strength in the jets is not known. Only various estimates are available: the field in the magnetosphere of a black hole is expected to be ~10^4 G (Field and Rogers, 1993; Krolik, 1999; Rees, 1984), though the estimates by Kardashev (1995) and by Shatsky and Kardashev (2002) suggest it could reach ~10^{11} G; the fields at distances <<1 pc from the black hole could be ~10^3 G (Rees, 1987); jet fields of ~0.01 G and ~100 G are considered in (Sikora et al., 1997); the field in the funnel along which the jet propagates is estimated to be <<700 G (Bednarek, 1993). In our model, the field strength in the jet is an unknown parameter. We obtain an estimate of the field based on the conditions for maximum acceleration of CRs in shock fronts in the jet.

What is chemical compositions of the accelerated CRs? Since we assume that the jet contains material from the accretion disk, the composition of the CRs should reflect that of the disk. The chemical composition of AGNs was investigated in the paper (Alloin et al., 2000) and a number of works referenced in it. We will assume that both protons and nuclei are present in the jet.

3.3. The Model for Acceleration of CR in Seyfert Nuclei

The main assumptions of our model are the following. In most Seyfert galaxies, relativistic jets form in the vicinity of a massive central black hole. (It is possible that they are disrupted at distances of 1– 3 pc inside a massive stellar kern.) The transverse cross section of the jet is $S = 3 \cdot 10^{31}$ cm^2, and the jet Lorentz factor is $\gamma = 10$ (Vilkoviskij and Karpova, 1996). Both protons and nuclei are present in the jets. The jet magnetic field is treated as an unknown, and we will obtain an estimate of the field based on the conditions for maximum acceleration of CRs in shock fronts in the jet.

The jet field is parallel to the jet axis, the shocks are also parallel. Particles are scattered by inhomogeneities in the magnetic field which are caused by turbulence, and are accelerated in parallel shocks to energies

$$E_j \approx Ze\beta_j B R_j \text{ erg},\qquad(4)$$

where Ze is the charge of the particle, β_j is the ratio of the jet velocity to the velocity of light, B is the magnetic field in a hot spot in the jet, and R_j is its transverse size (Norman et al., 1995). For the jet parameters presented above, the velocity and transverse size of the jet will be $\beta_j \approx 0.99$ and $R_j \approx 3 \cdot 10^{15}$ cm and the maximum particle energy will be

$$E_j \approx 1.9 \cdot 10^{18} ZB \text{ eV}.\qquad(5)$$

In the hot-spot magnetic field, the particle simultaneously is accelerated and loses energy to synchrotron radiation. We will assume that, under these conditions, the particle can accumulate the maximum energy if it loses less than half its acquired energy to synchrotron radiation over the acceleration time T_a. In other words, we assume that the acceleration time T_a does not exceed the time T_s over which the particle energy decreases by a factor of two: $T_a \leq T_s$. The acceleration time is equal to (Cesarsky, 1992)

$$T_a \approx l/(\beta_s^2 c) \text{ s,} \qquad (6)$$

where l is the diffusion mean free path (in the vicinity of a shock, this will be equal to the Larmor radius of the particle, $l \approx r_B$) and β_s is the ratio of the velocity of the shock to the velocity of light. The Larmor radius of a particle is given by (1), the time T_s is equal to (Ginzburg and Syrovatskii, 1964)

$$Ts \approx 3.2 \cdot 10^{18} / B_\perp^2 (A/Z)^3 1/Z (Mc^2/E) \text{ s,} \qquad (7)$$

where $B_\perp$ is the field component perpendicular to the particle's velocity, A is the atomic number of the particle, $M = Am_p$ is the mass of the particle, and m_p is the proton mass. Assuming $T_a \approx T_s$ and $E = E_{max} \approx 1/2 E_j$, we obtain from (4)–(7) the field strength for which the particle is accelerated to the maximum energy,

$$B_{CR} \approx 49.7 \, Z^{1/3} \text{ G,} \qquad (8)$$

and the values of this maximum energy for nuclei ($A/Z \approx 2$):

$$E_{max\,A} \approx 6.6 \cdot 10^{20} (Z/B)^{1/2} \text{ eV,} \qquad (9)$$

and for protons:

$$E_{max\,p} \approx 1.65 \cdot 10^{20} B^{-1/2} \text{ eV.} \qquad (10)$$

(These same formulas were obtained in a somewhat different way in our paper (Uryson, 2001c).) If the field is $B \sim (5–40)$ G, nuclei with $Z \geq 10$ acquire energies $E \geq 2 \cdot 10^{20}$ eV, while lighter nuclei are accelerated only to energies $E \leq 10^{20}$ eV: for protons, $B_p \approx 19.6$ G and $E_{max\,p} \approx 3.7 \cdot 10^{19}$ eV; for He nuclei, $B_{He} \approx 39.5$ G and $E_{max\,He} \approx 1.5 \cdot 10^{20}$ eV; and for Fe nuclei, $B_{Fe} \approx 16$ G and $E_{max\,Fe} \approx 8 \cdot 10^{20}$ eV. In a field of $B \sim 100$ G, particles with $Z > 2$ are accelerated to energies $E \geq 10^{20}$ eV. In a field of $B \sim 1000$ G, only heavy particles with $Z \geq 23$ acquire energies $E \geq 10^{20}$ eV.

The particles acquire such energies over distances of $R \approx T_a c$ from the base of the jet. Taking the values of B from above, we find that this distance is $R_{He} \approx 6 \cdot 10^{15}$ cm $\approx 20 R_g$ for He nuclei and $R_{Fe} \approx 3 \cdot 10^{15}$ cm $\approx 10 R_g$ for Fe nuclei, where R_g is the gravitational radius of a black hole with a mass of $10^9 M_O$. In a field of $B \approx 1000$ G, particles with $Z = 23$ are accelerated to $E = 10^{20}$ eV over a distance of $R \approx 0.1 R_g$.

Formulas (8)–(10) and these numerical estimates were obtained for a jet with the radius of $R_j \approx 3 \cdot 10^{15}$ cm. In a jet with radius kR_j, the maximum particle energy changes by only a factor of $k^{1/3}$. Therefore, these estimates are valid for jets with radii in the range $\sim 3 \cdot 10^{14}$–$3 \cdot 10^{16}$ cm.

Thus, protons are accelerated to $E < 4 \cdot 10^{19}$ eV, and do not fall into the UHE range for any values of B. Therefore, if this model is correct, CR protons with energies $E > 4 \cdot 10^{19}$ eV were not accelerated in Seyfert nuclei. They are fragments of nuclei, or were accelerated in other sources (such as BL Lac's). Further, the magnetic fields in jets can be estimated not only based on astronomical observations but also based on the energy spectrum and chemical composition of CRs.

3.4. Escape of the Particles from Their Sources

The accelerated particles lose energy in reactions with infrared photons and in processes of synchrotron and curvature radiation, which occure in magnetic fields. Here we consider these energy losses. (We do not consider particles interactions with the waves excited in the hot gas flow.)

3.4.1. Energy Losses of the Particles in Interactions with Photons

The particles lose energy in reactions with infrared photons, and the main energy losses occur in photopion reactions, $p + \gamma \rightarrow p(n) + \pi$, and in direct pair production reactions, $p+\gamma \rightarrow p+e^+$ $+e$. In the case of nuclei with masses A, photopion reactions give rise to the formation of m nucleons and n pions, $A + \gamma \rightarrow A' + mN + n\pi$. In addition, photosplitting of the nuclei can occur: $A + \gamma \rightarrow A' + mN$. The threshold energy (in the center of mass system) for photopion production is 145 MeV and for direct pair production is 1 MeV, so that the former reactions dominates at energies $E \geq 3 \cdot 10^{19}$ eV, while the latter dominates at particle energies $E \leq 2.1 \cdot 10^{18}$ eV, (Berezinsky et al., 1990). The pions produced in photoreactions decay via the paths $\pi^\pm \rightarrow$ $\mu^\pm + v$, $\mu^\pm \rightarrow e^\pm + v + v$, $\pi^0 \rightarrow 2\gamma$, and the electrons and photons that are produced in these decays can initiate cascades, leading to an increase in the radiation density and additional energy losses by the accelerated particles.

Losses are greatest in regions with high photon densities. Firstly, it is the funnel of the accretion disk. Secondly, it is the geometrically and optically thick dusty torus surrounding the central region of the galaxy (the thickness of the torus can reach ~100 pc) (Pier and Krolik, 1993). The torus radiates infrared photons, which are absorbed and reradiated by clouds inside the torus.

Energy losses in the funnel have been studied in (Bednarek and Kirk, 1995; Sikora et al., 1987). It was shown there that the particles lose only a negligible fraction of their energy in reactions of direct pair production if the radiation has a hard spectrum (as is the case for AGNs). Further, losses can be appreciable in photopion reactions, in which case collimated beams of gamma radiation form due to the development of cascades. Photpion losses are negligibly small in accretion disks with optical depths $\tau \leq 1$ (Kardashev, 1995; Norman et al., 1995). (Consequently, collimated beams of gamma radiation do not form in such disks.)

The passage of particles through the gas–dust torus is considered in (Norman et al., 1995). The particles lose an insignificant fraction of their energy in interactions with radiation rescattered in the torus if the luminosity of the galaxy is $L < 10^{46}$ erg/s.

In addition, a particle can lose energy in interactions with radiation that is emitted in the jet. The emission of such radiation is considered in (Falcke et al., 1995). Cascades can develop in the jets due to their interaction with the walls of the gas–dust torus. This enhances the radiation density in the jet, leading to additional energy losses by the accelerated particles. However, in Seyfert nuclei, the interaction of the jets with the torus walls is insignificant, so that no cascades develop. For this reason, these AGNs radiate only weakly in gamma rays.

An accelerated particle will not intersect the dust torus if it travels at an angle i to the normal to the galactic plane such that $\tan i < r/h$, where r is the inner radius of the torus and h is its thickness, i.e., if the galactic plane is tilted at a relatively small angle to the observer. The angle i is characterized by the ratio of the galactic semiaxes: $\cos i = e_2/e_1$ (if $e_2/e_1 = 0.6$, $i = 55^0$) (Simcoe et al., 1997). Therefore, the ratio e_2/e_1 should be comparatively large for CR sources.

As a result of interactions of the particles with infrared photons, electromagnetic cascades develop, giving rise to collimated X-ray radiation along with collimated beams of gamma rays (Stecker et al., 1991).

For these reasons, if the energy losses of the CRs within the funnel and dust torus are small, AGNs will have modest fluxes in X-ray and gamma-ray bands and will not have collimated beams of gamma and X-ray radiation.

3.4.2. Energy Losses in Magnetic Fields

The accelerated particles lose energy to synchrotron and curvature radiation in magnetic fields.

Synchrotron losses in the flow of gas are insignificant, since the field in these flows is directed (as in the jet) predominantly along the direction of motion. The curvature losses for a particle with charge Z are (Zheleznyakov, 1997)

$$-dE/dt = 2/3(Ze)^2 c(E/Mc^2)^4 (\rho_c)^{-2} \text{ erg/s} \tag{11}$$

where ρ_c is the radius of curvature of the field line. Hence, the particle energy halves over a time

$$T_{curv} = 7/2(Mc^2)^8 E^{-3}(Ze)^{-2}\rho_c^{\,2}c^{-1} \text{ s.} \tag{12}$$

The particle propagates along the field line a distance R_{line}. Particles with energy E_{max} travel this distance over a time

$$t \approx R_{line}/c \approx 4.6 \cdot 10^9 \text{ s.} \tag{13}$$

The curvature losses will be small if the particle loses no more than half its energy E_{max} as it moves along the field line:

$$T_{curv} > t. \tag{14}$$

We find the distance R_{line} as follows. The accelerated particle freely leaves the galaxy, having reached regions where the field has decreased enough that the Larmor radius of the particle becomes roughly $r_B \geq 5$ kpc (Pochepkin et al., 1995). (Host galaxies of most Seyferts are spirals, and we have taken the typical dimensions of the spirals to be the same as those of our Galaxy.) For ultrarelativistic particles with various values of Z and with energy $E = E_{max}$, the condition $r_L \geq 5$ kpc is fulfilled when the field is $B \leq 10^{-5}$ G. Taking the magnetic field to decrease with distance as $B \sim R^{-3}$ (private communication by Vilkoviskij), and the field at $R \sim 1$

pc to be $B \sim 1$ G (Rees, 1987), we obtain $R_{line} \approx 46$ pc. The curvature radius of the field lines in a dipole magnetic field is $\rho_c = 4R^2/3a$, where R and a are the distances from the center and from the axis of the dipole (Kardashev, 1995). Based on this information and (12)– (14), we can estimate the maximum deviation from the jet axis of a particle with energy $E = E_{max}$ that has traveled a distance $R \approx 46$ pc with small curvature losses: for protons, $a_p \approx 0.01$ pc; for He nuclei, $a_{He} \approx 0.03$ pc; and for Fe nuclei, $a_{Fe} \approx 0.04$ pc.

We determine now the fraction of accelerated particles that leave the source without significant curvature losses. For such particles, the deviation from the jet axis will be

$$\theta \leq a/R_{line} = 6.5 \cdot 10^{-4}. \tag{15}$$

In the wave system, the particles are scattered isotropically. Therefore, the fraction of particles in which we are interested can be found as follows. The angle θ^* between the velocity vector and jet axis in the wave system is related to the angle θ by the expression (Landau and Lifshits, 1990)

$$\tan \theta = (1 - \beta^2)^{1/2}(\beta + \cos \theta^*)^{-1} \sin \theta \tag{16}$$
$$= 0.14 \sin \theta^*(0.99 + \cos \theta^*)^{-1}.$$

with $\beta \approx 0.99$. For angles $\theta < 0.02$, we have $\sin \theta^* \approx \theta^*$, $\cos \theta^* \approx 1$ and $\theta \approx 0.07 \, \theta^*$. Hence, $\theta^* \approx 0.01$, and the fraction of particles deviating from the axis by angles (15) is $\eta = 0.01/\pi \approx 3 \cdot 10^{-3}$; i.e., roughly one in 300 of the accelerated particles leaves the source without curvature losses.

3.5. Estimates of the Energy of CR Sources

Based on the flux of CRs measured at the Earth, we can estimate the observed luminosities in UHECRs for Seyfert nuclei. This luminosity L_{CR} is equal to

$$L_{CR} = \int_E^\infty F_g(E)EdE, \tag{17}$$

where $F_g(E) = KE^{\gamma}$ is the differential spectrum of particle generation in a Seyfert nuclei. If the spectrum of UHECRs is weakly distorted by interactions in intergalactic space (this is possible if the sources are located within $R \sim 50$ Mpc of the Earth), the particle-generation spectrum in the source $F_g(E)$ and the observed CR spectrum $I(E)$ for $E > 5 \cdot 10^{19}$ eV will have the same form, and $F_g(E) = KE^{-\chi}$ with $\chi \geq 3$. The intensity of CRs with energy E in the Universe is (Berezinsky et al., 1990)

$$I(E) = (c/4\pi)F_g(E)n_S T_{CR}, \tag{18}$$

where $n_S = 2 \cdot 10^{-77}$ cm^{-3} is the density of Seyfert galaxies and T_{CR} is the age of the UHECRs. Since the distances from which the CRs can arrive without interacting with the background radiation are $R \sim 50$ Mpc, the age is $T_{CR} = R/c \approx 1.7 \cdot 10^8$ yr. According to measurements made at various installations (Watson, 1996), the spectrum $I(E)$ at $E > 5 \cdot 10^{19}$ eV is $I(E) \approx 10^{-39} - 10^{-40}$ (cm^2 s sr eV)$^{-1}$. We obtain from (17) and (18) the observed luminosity of the Seyfert nuclei in CR with $E > 5 \cdot 10^{19}$ eV, $(L_{CR})_S \approx 10^{41} - 10^{42}$ erg/s, $\chi \approx 3.1$. If $\chi \approx 3$ the luminosity is $(L_{CR})_S \approx 10^{39} - 10^{40}$ erg/s.

The real luminosity is higher than the observed value by a factor of $1/\eta \approx 300$. (Note that, if the mass of the black hole is $M \approx 10^9 M_O$ and all its energy Mc^2 is spent on accelerating UHECRs at the power found above, the energy of the black hole is depleted over $10^{13} - 10^{14}$ yr, much longer than the age of the Universe, $T_{Mg} \approx 1.3 \cdot 10^{10}$ yr. Therefore Seyferts have enough energy for CR acceleration.)

3.6. Results

Particles can be accelerated to ultrahigh energies $E \approx 10^{21}$ eV in Seyfert nuclei. This acceleration occurs in shock fronts in relativistic jets. The maximum energy and chemical composition of the accelerated particles depend on the magnetic field in the jet, which is not well known; fields in the range ~5–1000 G are considered in the model. The highest energies of $E \approx 10^{21}$ eV are acquired by Fe nuclei when the field in the jet is $B \approx 16$ G. When $B \approx (5-40)$ G, nuclei with $Z < 10$ are accelerated to $E \leq 10^{20}$ eV, while nuclei with $Z \geq 10$ acquire energies $E \geq 2 \cdot 10^{20}$ eV. Only particles with $Z \geq 23$ acquire energies $E \geq 10^{20}$ eV when $B \approx 1000$ G. Protons are accelerated to $E \approx 3.7 \cdot 10^{19}$ eV, and do not fall into the range of UHE for any magnetic field B.

The conditions in Seyferts allow the particles escape from sources with small energy losses. The particles lose a negligible amount of their energy in the accretion disk if its optical depth is $\tau \leq 1$; losses in the thick gas–dust torus are small if the luminosity of the galaxy is $L \leq 10^{46}$ erg/s and the angle between the normal to the galactic plane and the line of sight is suffciently small, i.e., if the axial ratio of the galactic disk is comparatively high. The particles do not lose energy to curvature radiation if their deviations from the jet axis do not exceed 0.03–0.04 pc at distances of $R \approx 40-50$ pc from the center. Synchrotron losses are small, if the magnetic field frozen in the galactic wind at $R \leq 40-50$ pc is directed (as in the jet) primarily in the direction of motion.

If the model considered is valid, the detected CR protons could be either fragments of Seyfert nuclei or be accelerated in BL Lac's. The jet magnetic fields can be estimated both from direct astronomical observations and from the energy spectrum and chemical composition of CRs.

4. The Maximum Energy and Spectra of Cosmic Rays Accelerated in AGNs

4.1. Introduction

In intergalactic space, particles interact with background radiation; as a result, they inevitably lose their energy (Greisen 1966; Zatsepin and Kuzmin 1966). Particles of different energies traverse different distances without significant energy losses. These distances for UHECRs were estimated in (Stecker, 1968; 1998).The redshifts $z \leq 0.0092$ of Seyfert nuclei correspond to distances up to 40 Mpc (for the Hubble constant $H = 75$ km s^{-1} Mpc^{-1}), in agreement with the results by Stecker (1968, 1998). The BL Lac's identified as possible CR sources are far, up to ~1000 Mpc (Veron-Cetty and Veron, 2001), away. Therefore, the question arises as to whether the particles accelerated in BL Lac's can reach an array with an energy of $3 \cdot 10^{20}$ eV (this is the maximum recorded CR energy (Bird et al., 1995)). In this Section, we compute the energy spectra of the incident (on an array) CRs in two simple cases: when CRs escape from AGNs with power-law and monoenergetic spectra. We compare the computed and measured spectra. For our calculations, we take the distribution of AGNs from the catalogue by Veron-Cetty and Veron (2001).

4.2. Description of the Model

According to the model by Kardashev (1995), particles in BL Lac's are accelerated in the electric field up to an energy of $10^{27}Z$ eV; the particle energy can decrease through curvature radiation to $10^{21}Z$ eV. Based on this acceleration mechanism, we assume in our calculations that the initial spectrum of the protons accelerated in BL Lac's is monoenergetic with an initial energy of 10^{27} and 10^{21} eV. Since particles in Seyfert nuclei maybe accelerated at shock fronts (Uryson, 2001c), we assume that the initial spectrum of the particles is a power law ($\sim E^{-\chi}$) with a spectral index of $\chi = 2.6$ and 3.0. Particles in Seyfert nuclei can be accelerated up to an energy of $8 \cdot 10^{20}$ eV.

The composition of the CRs with energies $E \approx 4 \cdot 10^{19} - 3 \cdot 10^{20}$ eV is not yet completely known. In accordance with the data of Shinozaki et al. (2005), we assume that the CRs with energies as high as 10^{21} eV are particles rather than gamma-ray photons.

The propagation of CRs in intergalactic space was considered under the following assumptions. The nuclei disintegrate into nucleons through their interactions with background radiation, traveling no more than 100 Mpc from their source (Puget et al., 1976; Stecker 1998). Therefore, if the CR sources are much farther than 100 Mpc, then, for simplicity, we may assume that the nuclei completely disintegrate near the source and consider only the propagation of protons in intergalactic space. Since the overwhelming majority of BL Lac's are $R > 400$ Mpc away (Veron-Cetty and Veron, 2001), this assumption is justified for the CRs emitted by BL Lac's. For simplicity, we assume that only protons propagate from Seyfert nuclei as well.

We computed the CR energy losses in intergalactic space under the following assumptions. Protons interact with relic and infrared photons. Protons with energies $E > 4 \cdot 10^{19}$ eV lose their energy mainly through the photopion reactions $p + \gamma \rightarrow N + \pi$; the losses through the electron–positron pair production are negligible (Blumenthal, 1970; Berezinsky et al., 1990). The density spectrum for relic photons with energy ε is described by the Planckian distribution

$$n(\varepsilon)d\varepsilon = \varepsilon^2 d\varepsilon / (\pi^2 h / (2\pi)^3 c^3 (exp(\varepsilon/kT) - 1)) \qquad (19)$$

with the temperature $T = 2.7$ K, the mean photon energy is $\langle \varepsilon \rangle \approx 6 \cdot 10^{-4}$ eV, and their mean density is $\langle n_0 \rangle \approx 400$ cm^{-3}. For the photons of the high energy tail in the Planckian distribution, the mean energy is $\langle \varepsilon_t \rangle \approx 1 \cdot 10^{-3}$ eV and the mean density is $\langle n_t \rangle \approx 42$ cm^{-3}.

The energy range of the infrared radiation is $2 \cdot 10^{-3}$–0.8 eV; at present, there are no detailed spectral measurements. We assumed that the infrared radiation spectrum is described by the numerical expression (Puget et al., 1976; Stecker, 1998)

$$n(\varepsilon) = 7 \cdot 10^{-5} \varepsilon^{-2.5} \text{ cm}^{-3} \text{ eV}^{-1}, \tag{20}$$

the mean energy of the infrared photons is $\langle \varepsilon_{IR} \rangle \approx 5.4 \cdot 10^{-3}$ eV, and their mean density is $\langle n_{IR} \rangle \approx 2.28$ cm^{-3}.

The photopion reactions are threshold ones. The threshold energy is $\varepsilon_{th}^* \approx 145$ MeV, where ε^* is the photon energy in the proton system; the threshold inelasticity coefficient is $K_{th} \approx 0.126$ (Stecker, 1968). The cross section σ and the inelasticity coefficient K of the photoprocesses depend on the energy ε^*. The dependences $\sigma(\varepsilon^*)$ and $K(\varepsilon^*)$ were taken from (Stecker, 1968) and (Particle Data Group, 2002). In addition to the photopion reactions, protons lose their energy through the expansion of the Universe. The adiabatic losses of a proton that propagates with an initial energy E from a point with a redshift z to a point with $z = 0$ are

$$-dE/dt = H(1 + z)^{3/2} E. \tag{21}$$

The cosmological evolution of the Universe was taken into account in the CR propagation. We used the Einstein–de Sitter model with $\Omega=1$, in which the time and the redshift are related by

$$t = 2/3 H^{-1} (1 + z)^{-3/2}; \tag{22}$$

the distance to an object at redshift z is

$$r = 2/3 cH^{-1}[1-(1 + z)^{-3/2}] \text{ Mpc.} \tag{23}$$

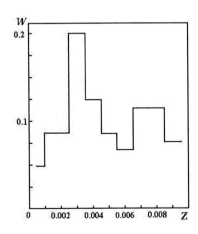

Figure 2. Redshift distribution of nearby ($z \le 0.0092$) Seyfert galactic nuclei normalized to the total number of objects.

At the epoch with a redshift z, the relic photon density and energy were, respectively, a factor of $(1 + z)^3$ and $(1 + z)$ higher than those at $z = 0$ (Berezinsky et al., 1990). We assumed that the particles propagate in the intergalactic magnetic fields almost rectilinearly. The sources of UHECRs, Seyfert nuclei at $z\leq0.0092$ and BL Lac's, were assumed to be distributed in redshift in accordance with the catalog by Veron-Cetty and Veron (2001). The z distributions of these objects at declinations $\delta\geq-15^0$ are shown in Figs. 2 and 3.

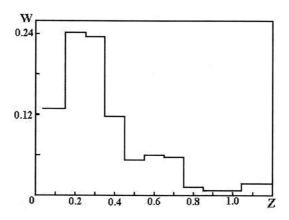

Figure 3. Redshift distribution of BL Lac objects normalized to the total number of objects.

4.3. Calculations

The calculations were performed as follows. First, we generated the redshift z_0 of a source by the Monte Carlo method in accordance with the distributions shown in Figs. 2 and 3. Subsequently, we calculated the distance to the source. Since the energy losses of the CRs depend on the distances that they traverse in intergalactic space, we determined distances by two methods to reach reliable conclusions. The first method uses formula (23). The second method assumes that $r = czH^{-1}$ (Mpc) for $z< 0.4$ (Pskovskii, 1990) and uses formula (23) for higher z. The calculations were performed with $H = 75$ and 100 km s^{-1} Mpc^{-1}. Next, we randomly generated the proton energy E and the angle θ in the laboratory system and determined the photon energy in the proton system

$$\varepsilon^* = \gamma\varepsilon(1-\beta \cos \theta), \tag{24}$$

where γ is the Lorentz factor of the proton, and $\beta = (1-1/\gamma^2)^{1/2}$. If $\varepsilon^* < \varepsilon^*_{th}$, then the proton interacted with photons of the high-energy tail in the Planckian distribution. If, alternatively, the photon energy ε^* was also below its threshold value in this case, then the proton interacted with infrared photons. The cross section σ and the inelasticity coefficient K for this interaction were determined from the value of ε^* (for details see (Uryson, 2004a)). Subsequently, we calculated the proton mean free path $<\lambda> = (<n>\sigma)^{-1}$, where $<n>=<n_0>$, $<n_t>$ or $<n_{IR}>$, depending on which photon the proton interacted with. Next, we generated the proton mean free path L by the Monte Carlo method and calculated the redshift z_1 of the proton after it traversed the distance L. At the point with z_1, the proton energy decreased due to its interaction with the photon by $(\Delta E)_{ph} = EK$. The decrease in energy due to the adiabatic losses at the point with z_1 is

$$(\Delta E)_{ad} = E(z_0 - z_1)/(1 + z_0) \tag{25}$$

This procedure was then repeated. The adiabatic losses at the point with redshift z_2 were got in the following manner. The point of the preceding interaction with redshift z_1 in formula (25) was taken in place of the point with z_0, the point with z_2 was taken in place of the point with z_1, an so on. The procedure ended if the proton reached the Earth (the point with $z_i = 0$) or if its energy decreased to $E < 4 \cdot 10^{19}$ eV.

4.4. Results

4.4.1. The Maximum Particle Energy in a Source

Aharonian *et al.* (2002) and Medvedev (2003) theoretically estimated the maximum CR energy in sources to be $\sim 10^{21}$ eV. The initial proton energy in BL Lac's without and with the inclusion of curvature losses in the source is, respectively, 10^{27} and 10^{21} eV (Kardashev, 1995). These estimates can be easily compared with the CR data by calculating the mean energies of the protons at the Earth, the protons having initial energies of 10^{27} and 10^{21} eV. We computed the energies of protons at the Earth using the Monte Carlo method assuming distribution of BL Lac's as in Figure 3. In each case, the statistics of simulated protons was 10^4. The mean proton energies on Earth were found to be 10^{24} and $6 \cdot 10^{19}$ eV, respectively. The first value is in conflict with the experimental data (recall that based on the possible CR acceleration mechanism (Kardashev, 1995), we assumed the initial CR spectrum in BL Lac's to be monoenergetic), the calculation with an initial energy of 10^{21} eV is consistent with the measurements, in agreement with the theoretical estimate above. This value is close to the maximum energy, $8 \cdot 10^{20}$ eV, of the particles emitted by Seyfert nuclei (Uryson 2001c, 2004b). For the subsequent analysis, let us consider the proton spectra at the Earth.

4.4.2. The Spectra of the Protons Incident on an Array

The measured CR spectrum at $E > 4 \cdot 10^{19}$ eV exhibits a flat component and a bump that are probably attributable to the CR energy losses in intergalactic space: these losses lead to the "transfer" of particles to the range of lower energies provided that the energy losses decrease with decreasing energy (Hillas 1968; Hill and Shramm 1985). Berezinsky and Grigor'eva (1988), Berezinsky et al. (1989), and Yoshida and Teshima (1992) computed the CR spectrum and analyzed its shape. The closer the source, the higher the energy at which a bump appears in the spectrum. We analyzed the shape of the measured spectrum in the energy range 10^{18}–10^{20} eV in the paper (Uryson, 1997).

Since the energies of the particles that trigger air showers are measured by different methods, the CR spectra measured on different arrays differ in intensity. The combined spectra normalized using measurements on a particular array are given in the literature. Here, we compare the computed spectra with published measurements.

The differential UHECR spectra measured on different arrays are shown in Figure 4: the spectra obtained on different arrays and normalized using the Fly's Eye array are shown in Figures 4a (Bahcall and Waxman, 2003) and 4d (Nagano and Watson, 2000); spectra obtained on the Pierre Auger Observatory and on the HiRes array are shown in Figures 4b, 4c (Yamamoto, 2007; Bergman, 2007). The computed spectra normalized using measurements at $E \approx 5 \cdot 10^{19}$ eV are shown in the same figures. The simulated proton statistic is 10^5 for each curve.

Let us first consider the spectra in Figure 4a. The large measurement errors make it difficult to compare the computed curves with the experimental data. However, the model with a monoenergetic initial CR spectrum in Seyferts is clearly inconsistent with the measurements. The models with a monoenergetic initial CR spectrum in BL Lac's and with a power-law initial CR spectrum in Seyfert's are suitable for describing the data of arrays except for the HiRes data points at $E < 10^{20}$ eV. The HiRes data are best described by the model with a power-law CR spectrum in BL Lac's at $\chi = 2.0$ and by the model with a power-law CR spectrum in Seyferts at $\chi = 3$. This is consistent with the possible particle acceleration conditions in these sources.

In Figures 4b, 4c, comparing the measured spectra and calculated curves, we conclude that initial spectra in sources are power-law, but the initial spectral index, 3.0 or 2.6, is difficult to determine due to large errors.

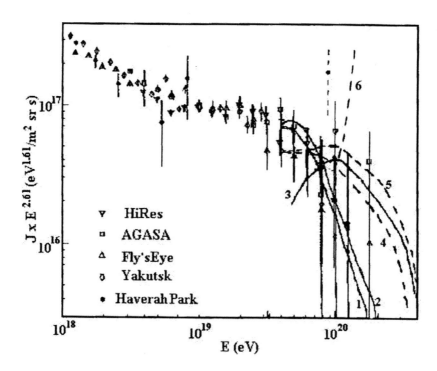

Figure 4. (a) Differential CR energy spectra as measured on different arrays from (Nagano and Watson, 2000). The curves represent the spectra computed for $H = 75$ km s^{-1} Mpc^{-1} where the distances were determined by the first method; the solid lines indicate the spectra of the CRs arrived from BL Lac's: a power-law initial spectrum in the sources with $\chi = 2.6$ (1), a power-law initial spectrum with $\chi = 2.0$ (2), and a monoenergetic initial spectrum (3); the dashed lines indicate the spectra of the CRs arrived from Seyfert nuclei: a power-law initial spectrum with $\chi = 3.0$ (4), a power-law initial spectrum with $\chi = 2.6$ (5), and a monoenergetic initial spectrum (6).

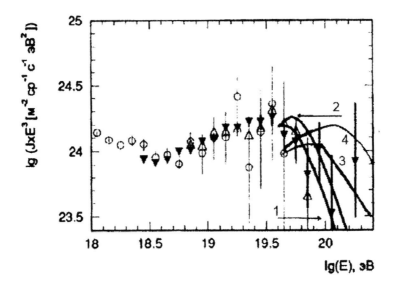

Figure 4. (b) The differential CR energy spectra as measured on the Pierre Auger Observatory from (Yamamoto, 2007) using showers at zenith angles above 60^0 (filled triangles) and below 60^0 (opened triangles) together with the spectrum derived from the hybrid data set (sircles). The solid lines indicate the calculated spectra of the CRs having power-law initial spectrum in the sources: CRs arrived from BL Lac's with the initial spectral index $\chi= 3.0$ (1), arrived from BL Lac's with $\chi= 2.6$ (2), arrived from Seyferts with $\chi=3$ (3); arrived from Seyferts with $\chi=2.6$ (4). The spectra were computed in the same way as those in Figure 4a.

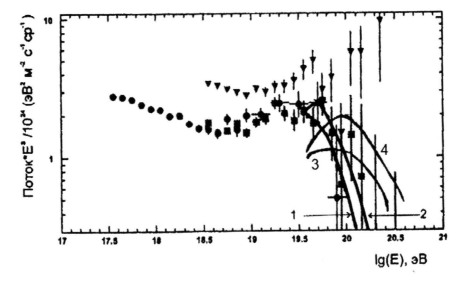

Figure 4. (c) The differential CR energy spectra as measured on the HiRes array from (Bergman, 2007) using HiRes-2 data set (sircles), HiRes-1 data set (squares), together with the spectrum derived from AGASA data set (triangles). The solid lines indicate the calculated spectra of the CRs having power-law initial spectrum in the sources: CRs arrived from BL Lac's with the initial spectral index $\chi= 3.0$ (1), CRs arrived from BL Lac's with $\chi= 2.6$ (2); CRs arrived from Seyferts with $\chi=3$ (3); CRs arrived from Seyferts with $\chi=2.6$ (4). The spectra were computed in the same way as those in Figure 4a.

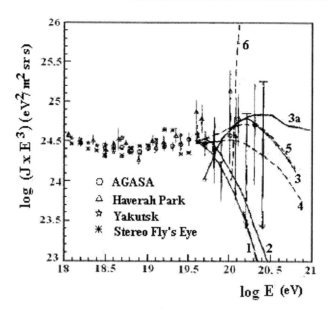

Figure 4. (d) The differential CR energy spectra as measured on different arrays from (Nagano and Watson, 2000). The spectra were computed for $H = 100$ km s^{-1} Mpc^{-1} where the distances were determined by the second method. The letter a marks the spectra of the CRs arrived from BL Lac objects for $z_{min} = 0.01$; in the remaining cases, the spectra from BL Lac's were computed for $z_{min} = 0.02$.

Figure 4d shows the spectra measured on different arrays, but the curves were computed for $H = 100$ km s^{-1} Mpc^{-1} where the distances were determined by the second method (see the section "Calculations"). In this case, the measurements are also described by the models with a monoenergetic initial CR spectrum in BL Lac's and with a power-law initial CR spectrum in Seyferts. However, the computed CR spectra in these models will differ greatly if 2 percent of the BL Lac's at redshifts $z < 0.1$ are assumed to be at the distance with $z = 0.01$ (according to the catalog by Veron-Cetty and Veron (2001), the minimum redshift for BL Lac's is $z = 0.02$).

It follows from the above analysis that the models for both far and nearby sources account for the measured CR spectrum at energies $E > 4 \cdot 10^{19}$ eV. In addition, spectra measured on the Pierre Auger Observatory and on the HiRes array confirm the model in which CRs in the sources are accelerated with the initial power-law spectrum. The maximal energy in the sources is about 10^{21} eV.

At energies below 10^{19} eV, the spectrum may be shaped by particles from distant sources (Berezinsky et al., 1990; Yoshida and Teshima 1992). According to the currently available data (Veron-Cetty and Veron 2001), the total number of Seyferts and BL Lac's is several thousand and several hundred, respectively.

4.4.3. Estimates of the CR Luminosity for Sources

We estimated the CR luminosity of Seyfert galactic nuclei in section 3. Here we estimate the observed CR luminosity for BL Lac's (L_{CR})$_{BL\,LAC}$:

$$(L_{CR})_{BL\,LAC} = U_{CR}/(NT), \qquad (26)$$

where U_{CR} is the total energy of the CRs emitted by BL Lac's, N is the total number of BL Lac's, and T is the CR lifetime. We can determine U_{CR} from the energy balance equation:

$$U_{CR} = (U_{CR})_{measured} + (U_{CR})_{lost}, \qquad (27)$$

where $(U_{CR})_{measured}$ is the energy of the CRs that reached the array, and $(U_{CR})_{lost}$ is the CR energy that was lost during the CR propagation from the source to the array. The initial CR energy in the source is $E_0 = 10^{21}$ eV; the bulk of the CRs on the array have an energy of $E = 5 \cdot 10^{19}$ eV. Assuming that

$$(U_{CR})_{measured}/U_{CR} \approx E/E_0 \approx 0.05, \qquad (28)$$

we obtain

$$U_{CR} \approx 20(U_{CR})_{measured}. \qquad (29)$$

We define $(U_{CR})_{measured}$ as

$$(U_{CR})_{measured} = 4\pi/cV \int_E I(E)E dE, \qquad (30)$$

where $I(E)$ is the computed intensity of the CRs from BL Lac's, and V is the CR-filled volume. The integral in (30) is equal to 4 eV cm^{-2} s^{-1} sr^{-1}. Most of the BL Lac's have redshifts $z \leq 0.35$ (see Figure 3); i.e., they are $r \leq 1000$ Mpc away. The CRs emitted by these sources reach an array in a time $T \leq 2 \cdot 10^{17}$ s. Assuming that the CRs fill a sphere with a radius $r \approx 1000$ Mpc and reach an array in a time $T \approx 2 \cdot 10^{17}$ s, we find that the total power of the sources is $U_{CR}/T \approx 2 \cdot 10^{44}$ erg/s. The number of sources at redshifts $z \leq 0.35$ is $N \approx 100$ (Veron-Cetty and Veron 2001). Therefore, the CR luminosity of a single source is $(L_{CR})_{BL\ LAC} \approx 2 \cdot 10^{42}$ erg/s. (The number of BL Lac's may be much larger; the luminosity $(L_{CR})_{BL\ LAC}$ is then lower than the value obtained above.) The power spent on the CR acceleration in a source is higher than its observed value, and is equal to $2 \cdot 10^{48}$ erg/s. It is because we assume in our estimates that the initial particle energy is 10^{21} eV, while these particles are accelerated in the source up to 10^{27} eV. According to the model by Kardashev (1995), CRs emerge from a source with an energy of 10^{21} eV due to the curvature losses, and the bulk of the energy is spent on gamma-ray radiation.

4.4.4. Discussion

The maximum CR energy is 10^{21} eV, irrespective of where they were accelerated, in Seyfert nuclei or in BL Lac's. This energy is close to the values obtained in the models by Haswell et al. (1992), Berezinsky et al. (1997), and Totani (1998): $\sim 10^{21}$ eV for the CRs accelerated in an accretion disk around a black hole with a mass of $\sim 10^7 M_O$, $\sim 3 \cdot 10^{21}$ eV if the particles are produced in the decays of metastable superheavy particles of cold dark matter, and $\sim 10^{21}$ eV if the CRs are accelerated in gamma-ray bursts. The maximum accelerated particle energy of 10^{21} eV was also obtained by Aharonian et al. (2002) and Medvedev (2003). The value of $\sim 10^{20}$ eV predicted in the model by Kichigin (2003) appears to be incorrect. In this model,

CRs are accelerated in the galactic magnetic fields by a surfatron mechanism. However, particles at energies of 10^{19} eV are not confined by the galactic magnetic fields, and their capture by suitable (for the subsequent surfatron acceleration) shock waves probably becomes impossible.

5. Conclusions

We here listed main results obtained in our model.

1. UHECR sources are AGNs, namely Seyfert nuclei with redshifts $z \leq 0.0092$, i.e. located at distances less than 40 Mpc, and BL Lac's. This result is got assuming extragalactic magnetic fields to be weak outside clusters of galaxies, $B < 10^{-9}$ G. Bl Lac's were identified as possible UHECR sources also by Tinyakov and Tkachev (2001).

2. Particles can be accelerated to ultrahigh energies $E \approx 10^{21}$ eV in the Seyfert nuclei. The acceleration takes place in shock fronts in relativistic jets in these objects. The maximum energy and chemical composition of the accelerated particles depend on the magnetic field in the jet, that is not well known at present. We considered fields in the range ~5–1000 G. The highest energy $E \approx 8 \cdot 10^{20}$ eV is acquired by Fe nuclei when the field in the jet is $B \approx 16$ G. When the field is $B \sim (5–40)$ G, nuclei with $Z \geq 10$ acquire energies $E \geq 2 \cdot 10^{20}$ eV, while lighter nuclei are accelerated to energies $E \leq 10^{20}$ eV. In a field of $B \sim 1000$ G, only particles with $Z \geq 23$ acquire energies $E \geq 10^{20}$ eV. Protons are accelerated to $E \leq 4 \cdot 10^{19}$ eV for any value of B. The estimates we have obtained are valid for relativistic jets with cross sections of ~$5 \cdot 10^{29}$–10^{33} cm^2. The particle acceleration occurs over distances ~$(0.1–10)R_g$. Losses in magnetic fields are small if CR deviations from the jet axis do not exceed $a \leq 0.03–0.04$ pc at distances $R \sim$ 40–50 pc, and if the magnetic field frozen in the galactic wind at $R \leq 40–50$ pc is oriented (as in the jet) predominantly along the direction of motion.

The Seyfert nuclei which were identified as possible UHECR sources satisfy conditions for escape of particles. These conditions are the following. The luminosity of the galaxy is $L < 10^{46}$ erg/s, and the angle between the normal to the galactic plane and the line of sight is sufficiently small, i.e., if the axial ratio of the galactic disk is comparatively high. These are conditions for small energy losses in interactions with photons.

The observed value of power emitted in UHECRs by a Seyfert nucleus is $(L_{CR})_S \approx 10^{40}$, 10^{42} erg/s, if the spectrum of CR generation in the source is exponential with the index $\chi = 3$, 3.1, respectively. The observed value of power radiated in UHECRs by a BL Lac is $(L_{CR})_{BL\ LAC} \approx 10^{42}$ erg/s. The power spent on CR acceleration in a Seyfert is higher by the factor of ~300, and in a BL Lac it is equal to $2 \cdot 10^{48}$ erg/s.

The model for CR acceleration in BL Lac's was suggested in (Kardashev, 1995).

3. The UHECR energy spectra at the Earth provide us with the information about the initial CR spectra in the sources. Data set of the Pierre Auger Observatory and Hires array indicates that initial spectra in sources are power-law, but the initial spectral index, 3.0 or 2.6, is difficult to determine due to large errors.

The maximal energy of CR at the Earth is of 10^{21} eV, no matter where they are accelerated, in Seyferts or in BL Lac's.

These are conclusions that we got in our model.

According to (Hillas, 1984; Aharonian et al., 2002) AGNs satisfy the general requirements for the CR sources which follow from classical electrodynamics.

Possible mechanisms for producing UHECRs are suggested in literature: in the evolution of topological defects (Berezinsky and Vilenkin, 1997), in decays of relic superheavy particles of cold dark matter (Kuz'min and Rubakov, 1998), in gamma-ray bursts (Totani, 1998). What do the models predict?

The maximal energy $\sim 10^{21}$ eV that we got is close to the values obtained in the models above and in the model for the CRs acceleration in an accretion disk around a black hole with a mass of $\sim 10^7 M_O$ (Haswell et al.,1992).

The following characteristics of UHECRs are expected in different models.. In the model of topological defects, most CRs with energies $E \approx 10^{21}$ eV should be gamma rays. In the model of decays of relic particles of cold dark matter, there should be an appreciable ($\sim 20\%$) excess of UHECRs from the Galactic center. Only UHE protons are produced in gamma-ray bursts.

Our model predicts: 1)no CR excess from the Galactic center, 2)UHECRs from Seyferts should be nuclei (nuclear fragments), 3)detected UHE protons are either nuclear fragments or were accelerated in BL Lac's.

If our model is correct, the jet magnetic fields can be estimated not only based on astronomical observations but also based on the energy spectrum and chemical composition of the CRs.

The model can be verified using measurements of the spectrum and composition of UHECRs. Such measurements will be carried out using AGASA, as well as using giant installations such as HiRes, Pierre Auger, Telescope Array, and satellite instruments (see references in the reviews by Bhattacharjee and Sigl (2000) and by (Nagano and Watson, 2000) and Proceedings of International Cosmic Ray Conferences.

Acknowledgements

I would like to thank V.L. Ginzburg, N.S. Kardashev, A.A. Starobinsky, V.S. Berezinsky, V.A. Dogiel, Ya.N. Istomin, B.V. Komberg, I.G. Mitrofanov, A.I. Nikishov, I.L. Rozental, O.K. Sil'chenko, V.A. Tsarev, Yu. N. Vetukhnovskaya, E.Ya. Vilkoviskij, and A.V. Zasov for discussions.

References

[1]   Afanasiev B. N. et al. 1996, in *International Symposium on Extremely High Energy Cosmic Rays: Astrophysics and Future Observatories,* ed. by M. Nagano. (Institute for Cosmic Ray Research. Tokyo), p. 32

[2]   Aharonian F. A. et al. 2002, *Phys. Rev.* **D 66**, 023005

[3]   Antonucci R. 1993, *Ann. Rev. Astron. Astrophys.* **31**, 473

[4]   Bahcall J. N., Waxman E. 2003, *Phys. Lett.* **B 556**, 1

[5]   Bednarek W.1993, *Astrophys. J.* **402**, L29

[6]   Bednarek W., Kirk J.G. 1995, *Astron.Astrophys.* **294**, 366

[7]   Begelman M.C., Cioffi D.F. 1989, *Astrophys.J.* **345**, L21

[8]   Begelman M.C. et al. 1984, *Rev. Mod. Phys.* **56**, 255

[9]   Berezinsky V. S., Grigor'eva S. I. 1988, *Zh. Eksp. Teor. Fiz.* **93**, 812

[10]   Berezinsky V. S. et al., 1989, *Zh. Eksp. Teor. Fiz.* **96**, 798

[11]   Berezinsky V. S. et al., in *Astrophysics of Cosmic Rays*, ed. by V.L.Ginzburg (North-Holland, New York, 1990).

[12]   Berezinsky V.S. et al. 1997, *Phys. Rev. Lett.* **79**, 4302

[13]   Berezinsky V.S., Vilenkin A., 1997, *Phys. Rev. Lett.* **79**, 5202

[14]   Bergman D.R. 2007, Proc. the 30th Int. Cosmic Ray Conf., Merida, Mexico

[15]   Bhattacharjee P., Sigl G. 2000, *Phys. Rep.* **327**, 109

[16]   Bird D. et al. 1995, *Astrophys. J.* **441**, 144

[17]   Blandford R., Eichler D. 1987, *Phys. Rep.* **154**, 1

[18]   Blumenthal G. R. 1970, *Phys. Rev.* **1D**, 1596

[19]   Bridle A.H., Perley R.A. 1984, *Ann.Rev .Ast ron. Astrophys.* **22**, 319

[20]   Cesarsky C. 1992, *Nucl.Phys. B (Proc. Suppl.)* **28**, 51

[21]   Cesarsky C., Ptuskin V. 1993, in *Proc. 23rd Intern. CR Conference, Calgary* **2**, 341

[22]   Chakrabarti S.K. 1988, *Mon. Not.R.Astron. Soc.* **235**, 33

[23]   Chechin V. A. et al. 2002, *Nucl. Phys. B (Proc. Suppl.)* **113**, 111

[24]   Cronin J.W. 1996, *International Symposium on Extremely High Energy Cosmic Rays: Astrophysics and Future Observatories*, ed. by M.Nagano (Institute for Cosmic Ray Research. Tokyo), p. 2

[25]   Falcke H. et al. 1995, *Astron. Astrophys.* **298**, 395

[26]   Falcke H. et al. 2000, *Astrophys. J.* **542**, 197

[27]   Fanaroff B.L., Riley J.M. 1974, *Mon. Not.R.Astron. Soc.* **167**, 31

[28]   Farrar G. R., Biermann P. L. 1998, *Phys. Rev. Lett.* **81**, 3579

[29]   Field G.B., Rogers R.D. 1993, *Astrophys. J.* **403**, 94

[30]   Forsythe G.E. et al. 1977, *Computer Methods for Mathematical Computations.* (N.-J. Prentice Hall, Inc., Englewood Cliffs, N.J.)

[31]   Gabuzda D.C., Cawthorne T.V. 2003, *Mon. Not.R.Astron. Soc.* **338**, 312

[32]   Ghisellini G., Celotti A. 2001, *Astron. Astrophys.* **379**, L1

[33]   Ginzburg V.L., Syrovatskii S.I. 1964, *The Origin of Cosmic Rays* (MacMillan, New York).

[34]   Gorbunov D.S. et al. 2002, *Astrophys. J.* **577**, L93

[35]   Greisen K. 1966, *Phys. Rev. Lett.* **16**, 748

[36]   Haswell C.A. et al. 1992, *Astrophys. J.* **401**, 495

[37]   Hayashida N. et al. 1996, *Phys. Rev. Lett.* **77**, 1000

[38]   Hayashida N. et al., 2000, *astro-ph* /0008102

[39]   Hill C. T. et al. 1987, *Phys. Rev.* **D 36**, 1007

[40]   Hillas A.M. 1984, *Ann.Rev. Astron. Astrophys.* **22**, 425

[41]   Hillas A.M. 1998, *Nature.* **395**, 15

[42]   Ho L.S., Peng C.Y. 2001, *Astrophys. J.* **555**, 650

[43]   Hoffman C.M. 1999, *Phys.Rev.Lett.* **83**, 2471

[44]   Hudson D.J. 1964, Statistics. Lectures on Elementary Statistics and Probability. Geneva

[45]   Istomin Ya.N., Pariev V.I. 1994, *Mon. Not. R. Astron. Soc.* **267**, 629

[46]   Kardashev N.S. 1995, *Mon. Not.R. Astron. Soc.* **276**, 515

[47]   Katsouleas T., Dawson J.M. 1983, *Phys.Rev. Lett.* **51**, 392

[48]   Kichigin N. G. 2003, *Phys. Dokl.* **48**, 159

[49]   Krolik J.H. 1999, *Astrophys. J.* **515**, L73

[50]   Kronberg P.P. 1994, *Rep. Progr. Physics.* **57**, 325

[51]   Krymsky G.F. 1977, Doklady of Academy of Sciences (DAN). **234**, 1306

[52]   Kuhr H. et al. 1981, *Astron. and Astrophys. Suppl. Ser.* **45**, 367

[53]   Kuzmin V. A., Rubakov V. A. 1998, *Yad. Fiz.* **61**, 1122 [*Phys. At. Nucl.* **61**, 1028]

[54]   Landau L.D., Lifshits E.M. 1990, *Field Theory* (Nauka, Moscow) [in Russian].

[55]   Lipovetsky V.A. et al. 1987, *Communications of the SAO*, No **55**

[56]   Mannheim K., Biermann P. L. 1992, *Astron. Astrophys.* **253**, L21

[57]   Medvedev M. V. 2003, *Phys. Rev.* **E 67**, 045401

[58]   Moderski R. et al. 1998, *Mon. Not. R. Ast ron. Soc.* **301**, 142

[59]   Nagar N.M. et al. 2001, *Astrophys. J.* **559**, L87

[60]   Norman C.A. et al. 1995, *Astrophys. J.* **454**, 60

[61]   Particle Data Group, *Phys. Rev.* **D 69**, 269 (2002).

[62]   Pier E.A., Krolik J. H. 1993, *Astrophys. J.* **418**, 673

[63]   Pochepkin D. N. et al. 1995, *Proc. 24th Int. Cosmic Ray Conf., Rome,* **3,** 136.

[64]   Popov S.B. 2000, http://xrai.sai.msu.su/polar/

[65]   Pskovskii Yu. P. 1990, *Space Physics,* Ed. By S. B. Pikel'ner (Sov. Entsiklopediya, Mosocow), p. 569 [in Russian]

[66]   Puget J. L. Et al. 1976, *Astrophys. J.* **205**, 638

[67]   Rachen J. et al.1993, *Astron. and Astrophys.* **273**, 377

[68]   Rees M.J. 1984, *Ann. Rev. Astron. Astrophys.* **22**, 471

[69]   Rees M.J. 1987, *Mon.Not.R. Astron.Soc.* **228**, 47 (1987).

[70]   Sagdeev R.Z., Shapiro V.U. 1973, *Zh. Eksp. Teor. Fiz. Lett.* **17**, 279

[71]   Shatsky A.A., Kardashev N.S. 2002, *Aston. Zh.* **79**, 708

[72]   Shinozaki K. et al. 2005, *Proc. the 29th Int. Cosmic Ray Conf., Pune,* HE1.4.

[73]   Sikora M. et al. 1987, *Astrophys. J.* **320**, L81

[74]   Sikora M. et al. 1994, *Astrophys. J.* **421**, 153

[75]   Sikora M. et al. 1997, *Astrophys. J.* **484**, 108

[76]   Sigl G. et al. 2001, *Phys. Rev.* **D63**, 081302(R)

[77]   Simcoe R. et al. 1997, *Astrophys. J.* **489**, 615

[78]   Spinrad H. et al. 1985, *Publ. Astron. Soc. Pac.* **97**, 932

[79]   Stanev T. et al. 1995, *Phys. Rev. Lett.* **75**, 3056

[80]   Stanev T. 1997, *Astrophys. J.* **479**, 290

[81]   Stecker F. W., 1968, *Phys. Rev. Lett.* **21**, 1016

[82]   Stecker F. W., 1998, *Phys. Rev. Lett.* **80**, 1816

[83]   Stecker F.W. et al. 1991, *Phys.Rev. Lett.* **66**, 2697

[84]   Squires G. L., 1968, *Practical Physics.* (McGraw-Hill, New York)

[85]   Takeda M. et al. 1999, *Astrophys. J.* **522**, 225

[86]   Thean A. et al. 2000, *Mon. Not. R.Astron. Soc.* **314**, 573

[87]   Tinyakov P.G., Tkachev I.I. 2001, *J. Exp. Theor. Phys Lett.* **74**, 445

[88]   Totani T. 1998, *Astrophys. J. Lett.* **502**, L13

[89]   Ulvestad J.S., Ho L.C. 2002, *Astrophys. J.* **562**, L133

[90]   Uryson A.V., 1996, *J. Exp. Theor. Phys Lett.* **64**, 77

[91]   Uryson A.V. 1997, *J.Exp.Theor.Phys. Lett.* **65**, 763

[92]   Uryson A.V., 1999, *J. Exp. Teor. Phys.* **89**, 597

[93]   Uryson A.V. 2001a, *Astron. Astrophys. Trans.* **20**, 347

[94]   Uryson A.V., 2001b, *Astronomy Rep.* **45**, 686

[95]   Uryson A.V., 2001c, Astron.Lett. **27**, 775

[96]   Uryson A.V., 2004a, *Astronomy Rep.* **48**, 81

[97]   Uryson A.V., 2004b, *Astron.Lett.* **30**, 816

[98]   Veron-Cetty M.P., Veron P. 1991, *ESO Scientific Report*, No **10**

[99]   Veron-Cetty M.P., Veron P. 1993, *ESO Scientific Report*, No **13**

[100] Veron-Cetty M.P., Veron P. 2001, *Astron. Astrophys.* **374**, 92

[101] Vilkoviskij E.Ya., Karpova O.G. 1996, Astron.Lett. **22**, 148

[102] Vilkoviskij E.Ya. et al. 1999, *Mon.Not. R. Astron. Soc.* **309**, 80

[103] Watson A., 1995, in *Particle and Nuclear Astrophysics in the Next Millenium*, eds. E.W. Kolb, R.D. Peccei. (World Scientific, Singapore), p. 126

[104] Watson A., in *Extremely High Energy Cosmic Rays: Astrophysics and Future Observatories*, ed. by M.Nagano (Inst. Cosmic-Ray Research, Tokyo, 1996), p.362.

[105] Wright G.S. et al. 1988, *Mon.Not.R. Astron. Soc.* **233**, 1

[106] Xu C. et al. 1999, *Astron. J.* **118**, 1169

[107] Yamamoto T. 2007, Proc. the 30th Int. Cosmic Ray Conf., Merida, Mexico

[108] Yoshida S., Teshima M. 1992, *Progr. Theor. Phys.* **89**, 833

[109] Zatsepin G.T., Kuz'min V.A. 1966, *J.Exp.Theor.Phys. Lett.* **4**, 78

[110] Zheleznyakov V.V. 1997, *Radiation in Astrophysical Plasma* (Yanus-K, Moscow) [in Russian].

INDEX

J

K

L

M

N